PUBLIC HEALTH IN THE 21ST CENTURY

ADVANCES IN HEALTH AND NATURAL SCIENCES

Public Health in the 21st Century

Additional books and e-books in this series can be found on Nova's website under the Series tab.

PUBLIC HEALTH IN THE 21ST CENTURY

ADVANCES IN HEALTH AND NATURAL SCIENCES

BURCU YUKSEL
AND
MUSTAFA SENCER KARAGUL
EDITORS

nova
Medicine & Health
New York

Library of Congress Cataloging-in-Publication Data

ISBN: 978-1-53614-639-4

Published by Nova Science Publishers, Inc. † New York

To Zeynep Esma, Zehra Ilgın and Mehmet Kürşad

CONTENTS

Introduction		**xi**
Chapter 1	The Spatial Analysis of Air Pollution Parameters (PM_{10}, SO_2 and NO_2): A Case Study of the Marmara Region (2014-2017) *Ozer Akyurek*	**1**
Chapter 2	Obesity, Obesity Surgery and Nursing Care *Züleyha Şimşek Yaban*	**17**
Chapter 3	Effect of Global Climate Change on Occupational Health: Heat Stress *Serap Arsal Yildirim*	**35**
Chapter 4	Physical Activity in Women *Asuman Saltan*	**51**
Chapter 5	The Enhanced Recovery after Surgery Protocol and the Role of Surgical Nurses *Selda Mert Boğa*	**65**
Chapter 6	Current Approaches to Prevent Diabetic Foot Ulcers *Ayfer Peker*	**83**

Chapter 7	A Review of Studies on the Application of Coumarin and Its Derivatives in Plants *Burcu Yuksel*	**97**
Chapter 8	Awareness of Brucellosis: A Worldwide Zoonosis *Mustafa Sencer Karagul*	**115**
Chapter 9	Mutagenic Potential of the Insecticide Arrivo 25 EC on Mitosis and Meiosis of *Hyacinthus Orientalis* *Ozlem Aksoy, Aylın Horozal and Burcu Yuksel*	**139**
Chapter 10	Aluminum-Induced Toxicity and Programmed Cell Death in Plants *Filiz Vardar, Fatma Yanık, Aslıhan Çetinbaş-Genç and Gamze Kurtuluş*	**155**
Chapter 11	Comparative Analysis of Different Regions of Rat Brain Tissue by FT-IR Spectroscopy *Sevgi Türker Kaya, Dilan Yıldırım, Didem Şentürk, Kübra Dağcı and Gül İlbay*	**183**
Chapter 12	Single Cell Gel Electrophoresis for Determination of Plant Responses to Genotoxic Stress *Yıldız Aydın, Ahu Altınkut Uncuoğlu and Filiz Vardar*	**203**
Chapter 13	Evaluation of the Sensitivity of the Nurses Working in Surgical Services of a Public Hospital to Patient Rights *Selda Mert Boga and Aylin Aydin Sayilan*	**223**
Chapter 14	Control and Eradication of Animal Brucellosis *Mustafa Sencer Karagul*	**243**
Chapter 15	Small Ruminant Morbillivirus *Ahmet Sait*	**261**

Chapter 16 Ultra-High Molecular Weight Polyethylene for Medical Applications **275**
Merve Dandan Doganci and Erdinc Doganci

Chapter 17 Nursing Care for the Symptoms of Cancer Treatment **299**
Aylin Aydin Sayilan

About the Editors **319**

Index **321**

INTRODUCTION

The edited volume focuses on most recent issues and topics in various fields of natural and health sciences. It includes chapters from important researchers and academics in the fields of agricultural, environmental, biological, medical sciences and biotechnology, among others. It reports recent research and includes the review of the most recent topics in these respective fields. In the field of environmental sciences, it includes chapters that focus on topics such as effects of air pollution in industrial cities of Turkey and effects of global climate change on various risk groups; in the field of nursing, the nursery support in obesity, the role of the nurses in the surgical process, the sensitivities of the nurses working in the surgical services and the impact of the nurses to overcome the side effects of cancer. It also covers topics about health such as Diabetes Mellitus, brucellosis and the effects of physical activity in women. Topics about biology and biotechnology such as Coumarin, Cypermethrin (Arrivo 25 EC), Aluminum toxicity and ultra-high molecular weight polyethylene (UHMWPE) are reviewed and researched by expert sciences. Moreover, chapters that utilized methods such as Fourier transform infrared (FT-IR) spectroscopy and COMET assay are also included in this book.

Researchers and graduate students need to keep up with the advances in natural and health sciences that occur almost daily. This edited collection of state-of-the-art reviews and top-notch research studies will provide scientists, educators, and researchers in this vital fields with the most recent developments and methods. This book, organized into seventeen chapters, features prominent scientists contributing to the current literature with their expertise in natural, environmental, biological and health sciences. The edited book will be of significant value to researchers, graduate students as well as practicing scientists working in these vibrant fields.

This volume is composed of the following parts:

In Chapter 1, Ozer Akyurek portrays the effects of air pollution in Marmara Region of Turkey. Air pollution is the increase of foreign matters in the air that affect living creatures' health in a negative way or cause physical injuries, above normal number and density. While air pollution-related effects and health problems are also seen at healthy people, especially elders and children in particular, air pollution creates serious health problems on human health and causes approximately more than 3 million people's death every year. In this chapter, the author provides spatial pollution distribution maps of Marmara Region that receives migration in high amount because of the intensive industrial activity in Turkey, which are created by the identification of air pollutant parameters' (PM_{10}, SO_2 and NO_2) patterns obtained by air quality monitoring stations of Marmara Clean Air Central and Ministry of Environment and Urbanization between the years of 2014-2107.

In Chapter 2, Zuleyha Simsek Yaban focuses on the nursery support in obesity. As we all know, obesity is not just an emerging health problem, it also leads to other public health problems, including: diabetes, hypertension, cardiovascular diseases, cancer, depression, infertility, and it negatively affects social life as well. In her chapter, she also states that the success of obesity surgery requires a multidisciplinary team approach including surgeons, psychiatrists, nutritionists, nurses, physiotherapists and other health professionals from the moment patients apply to hospital until being discharged. Providing personalized, individualized and holistic nursing care and counseling services to the patients who will undergo obesity surgery patients to reduce possible risks in their care, and preventing complications and difficulties before, during and after surgery are very crucial.

In Chapter 3, Serap Arsal Yildirim examines the effects of global climate change on various risk groups. She uses the data collected from the publications from the famous databases and websites such as "PubMed", "CDC" and "WHO" between June-July 2018 to present her findings. The study also emphasizes the effects of greenhouse gases emitted uncontrollably and intensely atmospheres with the Industrial Revolution on working health as a cause of global climate change.

In Chapter 4, Asuman Saltan investigates the effects of physical activity in women. In previous studies, it has been found that the tendency toward sedentary behaviors of women are higher than that of men. One reason for this is that women are still an influential value to society, culture, and the economy. Therefore, women and physical activity need to be examined even in today's conditions. This chapter includes evidence-based information on the implementation and effects of physical activity in women.

In Chapter 5, Selda Mert Boga emphasizes the importance of Enhanced Recovery after Surgery Protocol (ERAS). ERAS is a blood-based care model designed to accelerate post-operative healing and provide recommendations for patient care applications to be performed before, during and after the surgery. In a protocol that requires a multidisciplinary approach, surgical nurses play an active role at each stage of the process. The protocols were established to reduce the length of hospital stay of patients because the healing of patients has been negatively affected due to long hospitalization as a result of traditional practices. In this chapter, ERAS protocol components, recommendations and the role of the nurses in the surgical process are emphasized.

In Chapter 6, Ayfer Peker focuses on Diabetes Mellitus (DM). DM is a common metabolic disorder that can cause various complications. According to the studies, the prevalence of diabetes is rapidly increasing both in Turkey and in the World. It is an important health issue for the high healthcare costs and serious complications that may involve the patients with different aspects. In her chapter, she also states that treatment and care of the diabetics require a multidisciplinary approach because of the extensive nature of the problem. Therefore, regular diabetic neuropathy assessment, vascular assessment, joint and muscle examination, and walking analysis are recommended.

In Chapter 7, Burcu Yuksel summarizes the use of coumarin, a secondary metabolite, and its derivatives in different aspects of plant growth and formation. Coumarin and its derivatives have broad biochemical and pharmacological activities. They also have allelopathic effects mainly on plants. Summaries and review of previous research are presented in this chapter by pointing out significant contributions and effects of coumarin and its derivatives in plant growth and inhibition.

In Chapter 8, Mustafa Sencer Karagul focuses on brucellosis, which is one of the most common zoonotic diseases that causes abortion, infertility and reduced milk production in animals. Brucellosis in humans, on the other hand, is defined as an acute or sub-acute febrile illness including an intermittent or remittent fever. Because of brucellosis, many countries run the risk of suffering from economic and health problems like losing husbandry and infection risks. In his chapter, the author describes this disease and together with its species and biovars, explain its historical background, and reveal the ways it affects animals and humans.

In Chapter 9, Ozlem Aksoy and her colleagues explore the effects of Arrivo 25 EC (250g/l, Cypermethrin) insecticide applied to *Hyacinthus orientalis* grown in greenhouse at recommended dosage by manufacturing company and double the recommended dosage were investigated on mitosis in root tip cells and pollen meiosis. Based on their findings, they state that the insecticide used in trial applications caused anomalies in mitosis and pollen meiosis in parallel with the increase in dosage and this condition will have a negative effect on pollen fertility.

In Chapter 10, Filiz Vardar and her colleagues focus on Aluminum toxicity. It is one of the important abiotic stresses limiting plant growth. Aluminum (Al) is the third most abundant elements in the soil after oxygen and silicone; comprising approximately 7%. Whereas it exists primarily in the form of insoluble aluminosilicates or oxides, the complex Al becomes solubilized in acidic soils (< pH 5.0). After Al absorption by plant roots, the toxicity effects root growth adversely resulting in poor uptake of water and nutrients. Even if many researchers have performed to understand the mechanism of Al toxicity and tolerance, it has not yet been fully characterized because of the complicated chemical nature of Al. Moreover, the underlying molecular, genetic and physiological features are still not clear. Although this chapter describes Al toxicity and PCD principles generally, the authors focus especially on the mechanisms of Al-induced PCD which have been largely ignored with the goal in mind for environmental monitoring.

In Chapter 11, Sevgi Turker Kaya and her colleagues investigate the very complex system of brain in their research study. More specifically, they focus on Wistar rat whole brain and different brain regions such as right-left hemisphere, hippocampus and cerebellum were comparatively analyzed by Fourier transform infrared (FT-IR) spectroscopy. Within the scope of detailed spectral analysis, they analyze the frequency and band areas of some specific modes such as olefinic, CH_2 asymmetric and symmetric, amide I, C=O and PO^-_2 asymmetric stretchings were evaluated. In addition, the bandwidths of CH_2 asymmetric and symmetric bands. They also calculate olefinic =CH/ lipid, CH_2/lipid, lipid/protein ratios so that differences in parameters such as protein and lipid amounts, lipid packing and lipid fluidity can be obtained. As a result of their study, the authors conclude that FT-IR spectroscopy can be successfully applied for comparative analysis of distinct part of the same tissue.

In Chapter 12, Yıldız AYDIN and her colleagues, in their review study, give information about COMET assay and its applications as a valuable tool for monitoring DNA damage in different plant systems. They performed single cell gel electrophoresis studies to determine the genotoxic effects of different concentrations of pesticides, heavy metals and different plant growth regulators used in tissue culture applications in tobacco as model organisms, cereals (barley and wheat) and sunflower, respectively.

In Chapter 13, Selda Mert Boga and Aylin Aydin Sayilan, in their descriptive study, analyze the sensitivities of the nurses working in the surgical services of a state hospital on patient rights. Their participants were 54 nurses working in the surgical services of a state hospital between July 26 and October 1, 2010. The data of the study were collected with the personal information form and the Belief of Patient Rights Form (BPRF). In the statistical analysis of the results, they use frequency, percentage, mean, standard deviation, Mann Whitney U and Kruskal Wallis-H tests. Based on their results, they conclude that the surgical nurses who were 29 years old or younger, university graduates, had less than 5 years' study period and took courses on patient rights, had high sensitivity to patients' rights.

In Chapter 14, Mustafa Sencer Karagul provides general aspects of control and eradication of *Brucella* spp, which is one of the most common zoonotic diseases and which brings about important problems related to health and economy. It also gives information about the case in Turkey. As we all know, direct or indirect contact with infected animals or their products is the main way of transmission of the infection. Based on his review, the author states that if brucellosis in animal reservoirs can be controlled, it will be easier to see a decline in the incidence of human brucellosis as well.

In Chapter 15, Ahmet Sait focuses on small ruminant morbillivirus (SRM), also known as 'goat plague', 'Kata', 'syndrome of stomatitis-pneumoenteritis' or 'ovine rinderpest'. It is a viral disease of sheep, goats and small wildlife ruminants characterized by fever, sores in the mouth, diarrhea, pneumonia, and sometimes death. The chapter shares current information about the disease, which is considered to be a major barrier to the development of sustainable agriculture across the developing countries in the world.

In Chapter 16, Merve Dandan Doganci and Erdinc Doganci, in their review study, focus on ultra-high molecular weight polyethylene (UHMWPE). UHMWPE is a subset of the engineering thermoplastic polymers with extraordinary mechanical and physical properties with a molecular weight of 2–6x10^6 $gmol^{-1}$. It has a great complex of convenient properties including thermal stability, wear resistance, and high impact strength, compared to customary polyethylene attained at low pressure. Additionally, UHMWPE has exceptional sound dampening characteristics, excellent electrical insulating properties, superior dielectric, chemical resistance, and low moisture permeability. In this chapter, they also explain the morphological structure and properties of UHMWPE and its applications for the manufacturing of medical devices.

In Chapter 17, Aylin Aydin Sayilan reviews the impact of the nurses to overcome the side effects of cancer. As we all know, cancer is the most common disease nowadays and has lots of side effects for patients and nursing care intended for symptoms is very important. In her chapter, the author discusses side effects of the cancer and how nursing care can be incorporated to lessen these side effects.

September, 2018
Editors
Dr. Burcu Yuksel
Dr. Mustafa Sencer Karagul

In: Advances in Health and Natural Sciences ISBN: 978-1-53614-639-4
Editors: Burcu Yuksel et al.

Chapter 1

THE SPATIAL ANALYSIS OF AIR POLLUTION PARAMETERS (PM_{10}, SO_2 AND NO_2): A CASE STUDY OF THE MARMARA REGION (2014-2017)

Ozer Akyurek*
Geomatics Engineering Department,
Kocaeli University, Kocaeli, Turkey

ABSTRACT

Air pollution is the increase of foreign matters in the air above normal number and density. This affects the health living creatures in a negative way or may cause physical injuries, while air pollution-related effects and health problems can be also observed in healthy people, in especially elderly people and children, air pollution creates serious health problems on human health and causes more than 3 million people's death every year. This necessitates the examining and monitoring processes of air pollution and taking precautions against it, whenever needed. . In this study, spatial pollution distribution maps of the Marmara Region that

* Corresponding Author Email: ozerakyurek@gmail.com.

receives migration in high amount because of the intensive industrial activity in Turkey, are created by the identification of air pollutant parameters' (PM_{10}, SO_2 and NO_2) patterns which are obtained by air quality monitoring stations of Marmara Clean Air Central and Ministry of Environment and Urbanization between the years of 2014-17.

Keywords: air pollution, geostatistical analysis, spatial analysis, GIS

INTRODUCTION

Urbanization is defined as the increase in the urban population by the breakdown of the population between the cities and rural areas, and by the migration of people from rural areas to the cities. It is argued that within the period between the 1800s to these days, world population approximately doubled and this increase brought significant problems. While the living rate was 2,4% in the locations that have 20.000 and more population in the 1800s, this rate increased to 4,3% in the 1850s and to 9,2% in the first years of the 1990s. The highest increase took place in the 1950s as 20,9%. Along with the rapidly increasing population in a very short time period, pollution in cities showed a rapid increase (Demirarslan and Akıncı, 2018; Partigöc and Cubukcu, 2017).

Air pollution is the increase of foreign matters in the air above normal number and density. This pollution that affects the health of living creatures' in a negative way or cause physical injuries. . Air layer, by getting polluted with the waste that emerges as a result of human beings' various activities and shows up during production and consumption activities, affects the life of human beings on earth in a negative way (Akyurek et al., 2013). Global warming, industrial activities, consumption of fossil fuel that are used in motor vehicles, uncontrolled population growth and unplanned urbanization are listed as the primary factors that cause increase in pollution (Partigöc and Cubukcu, 2017).

Particulate matters (PMs) are the tiny granular solid or liquid particles that are spreading in the atmosphere and stay suspended in air. PMs may consist of several natural and artificial sources like burning of fuels, diesel engines, construction and industrial activities, secondary aerosols (reactions of ammoniac, Sulphur and nitrogen oxides in the air), herbal pollens and dusts that are raised from the ground. PMs cause serious health problems according to their sizes. As the sizes of PMs become smaller, they enter into the deepest points of lungs and destroy the structure and chemistry of lungs.

Sulphur dioxide (SO_2) is a colorless, suffocating and acidic gas released by the burning of the Sulphur compounds that are found in its structure. Main existing Sulphur dioxide producing activities are industrial applications, domestic fuels that are used for heating purposes, steam power plants and the use of diesel vehicles (Akyurek et al., 2013; Demirarslan and Akıncı, 2018).

Nitrogen dioxide (NO_2) is a reddish-brown, pungent gas. Primary sources of NO_2 are the fuels that are used in vehicles, industrial plants and domestic heating. Its stay in the atmosphere changes according to the height. It can stay several months at the sublayers of the atmosphere, at the upper layers, on the other hand, it can stay only a few days. It causes serious problems in respiratory tracts and it may be fatal if it is inhaled in high doses (Cetin, 2006).

It is stated in the previous studies that long exposure to the air pollution has detrimental effects on human health. Diseases like cough, bronchitis, heart disease and lung cancer show an increase depending on the air pollution. While air pollution-related effects and health problems are also seen in healthy people, especially elders and children, it creates serious health problems on human health. Therefore, examining, monitoring and taking precautions for air pollution is a significant issue. This problem causes more than 3 million people's death every year.

Geostatistics is an applied branch of statistics and it is a set of statistical methods which are evaluated with spatial data in earth sciences. Apart from the classical statistical methods, population parameters can be estimated related to the examples that are depended on one another instead of the examples that are independent of each other rule. Geostatistical analyses are used for predicting the value of the locations, in which observation is not carried out. In this sense, continuous surface format maps of data are obtained by making interpolation at the points that are not measured in the land (Akyurek et al., 2013). In literature, there are many studies that have investigated and analyzed air pollution via geostatistical methods (Elbir, 2004; Fedra and Haurie, 1999; Lin and Lin, 2002; David et al., 1997; Briggs et al., 2003; Chen et al., 2015; Hee-Jae and Myung-Jin, 2014; Xiao et al., 2014; Rohde and Muller, 2015; Matejicek, 2005; Song, 2008, Demirarslan and Akıncı, 2018; Kim et al., 2018, Moral et al., 2010, Esperanza et al., 2004, among others).

This study aims to create spatial pollution distribution maps of the Marmara Region, which receives migration in high amounts because of the intensive industrial activity in Turkey, by the identification of air pollutant parameters' (PM_{10}, SO_2 and NO_2) patterns which are obtained by air quality monitoring stations of Marmara Clean Air Central and Ministry of Environment and Urbanization between the years of 2014-17.

MATERIALS AND METHODS

Study area, the Marmara Region, is a region that is located between the 26° and 31° east longitudes, 38° and 42° north, hosts Turkey's cities which are developed industrially, and has 24,816,023 population according to the Turkish Statistical Institute's data and address-based population information. Increasing population and rapid industrialization bring along rural-urban migration movement. The Marmara Region, which is one of the important industrial zones of Turkey, and has been exposed to continuous movement of migration. It received 3,240,686 immigrants from the other cities between the years of 2014 and 2017, which is the period of study, according to the Turkish Statistical Institute's data. When the highest migration receiving cities are examined, cities like Istanbul, Bursa, Kocaeli and Tekirdag that have developed industrial activities, come to the forefront (Figure 1).

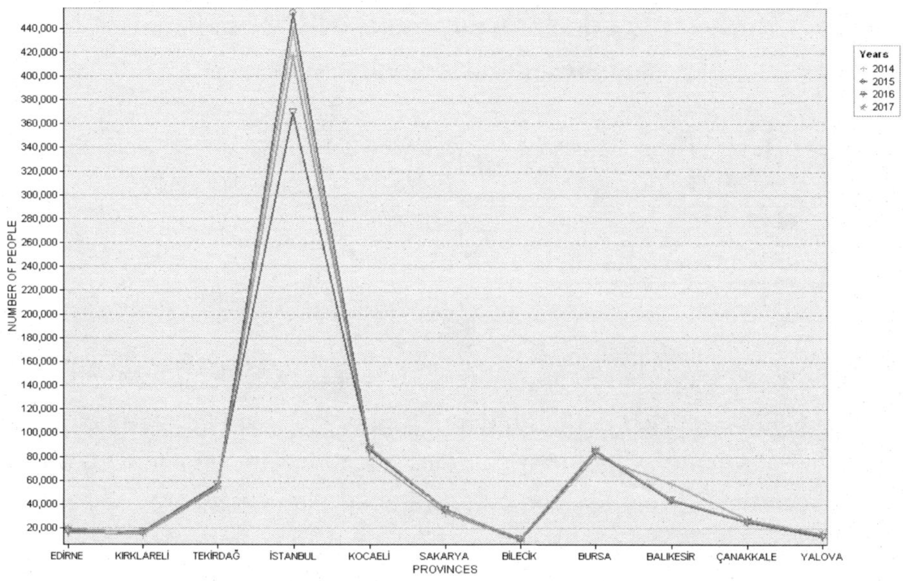

Figure 1. Migration distribution diagram according to the provinces.

The increase in the urban population brings along various environmental pollution issues. One of the most important pollutions is the air pollution. Air quality monitoring stations have been established throughout the country since 2005 in order to detect, analyze and take necessary precautions, in a timely fashion, for air pollution. Air quality is monitored with hourly measurements by 125 stations that are established by the Ministry of Environment and Urbanization in 81 provinces and 39 stations that are established by the Marmara Clean Air Center. In many countries, limit values, which are determined by the European Union and World Health Organization, are used as the local limit values in order to keep air pollution under control. Limit values for air pollutants in long term, short term and winter periods are determined in the regulations of Air Quality Evaluation and Management that was published in 2008 in Turkey. PM_{10} and SO_2 pollutants' limit values that belong to Turkish Limit Values, European Union and World Health Organization are shown in Table 1.

Table 1. PM_{10} and SO_2 pollutants with the national and international annual limits

Pollutant	Turkish Limit Values (AQEMR) (μg/m^3)	EU Regulations (μg/m^3)	WHO Guidelines (μg/m^3)
PM_{10}	44	40	20
SO_2	20	20	-
NO_2	44	40	40

Data used in this study is obtained by the 45 Air Quality Monitoring Stations that are located in the Marmara Region between the years of 2014 and2017. Air Quality Monitoring Stations' locations are shown in Figure 2. Hourly recorded PM_{10}, SO_2 and NO_2 pollutant parameters are used in the analyses conducted by the transformation of daily values first, and then monthly and yearly values.

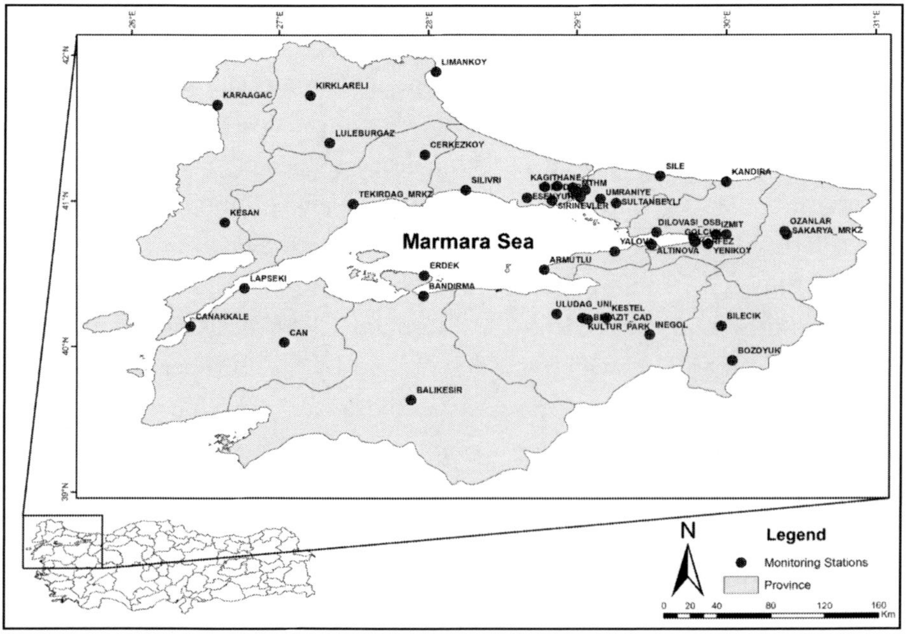

Figure 2. 45 Air quality monitoring stations available in the Marmara Region.

The problem of classical statistical methods is the assumption that data variables are independent of each other. It is not possible to obtain exact statistical independence for the values that are distributed in a specific structure. According to Tobler (1970), closer things are more related to each other than the further ones. The spatial dependence, which is defined as the observations that are close to each other are more similar than the observations that are remote to each other, invalidates classical statistics' independence assumption (Atkinson, 2002). Geostatistics is a statistical calculation method that approaches the relationship among the data by also taking account of their locations. The method has significant advantages like the detection of the error quantity at particular confidence levels. Geostatistical analyses are used for predicting the values of the locations, in which observation is not carried out. In this sense, continuous surface format maps of data are obtained by making interpolation at the points that are not measured in the land. These operations consist of two parts: modelling (semivariogram calculations) and spatial interpolation (Kriging method). The semivariogram is the calculation of the sample variance's change with the distance. Closer samples show similar variance value and generally show an increase with the distance between variance samples compared to the distant ones. The Kriging method, on the other hand, is used to predict features of spatial structure between the samples that are obtained by the semivariogram calculations, at the points that sampling is not done (Akyurek et al., 2013; Yesilkanat et al., 2014).

The semivariograms are specified for the study area, in which the measurement is made, as in the equation below;

$$\gamma(h) = \frac{1}{2N(h)} \sum_{i=1}^{N(h)} \left(Z(x_i) - Z(x_j + h) \right)^2 \qquad (1)$$

Here, $\gamma(h)$ demonstrates the semivariance value, the distance between h, i and j points, N(h) demonstrates the number of dot pairs at h distance, $Z(x_i)$ demonstrates the measurement value of data in the data set at the i point and $Z(x_j+h)$ demonstrates the measured value of the data in h distance.

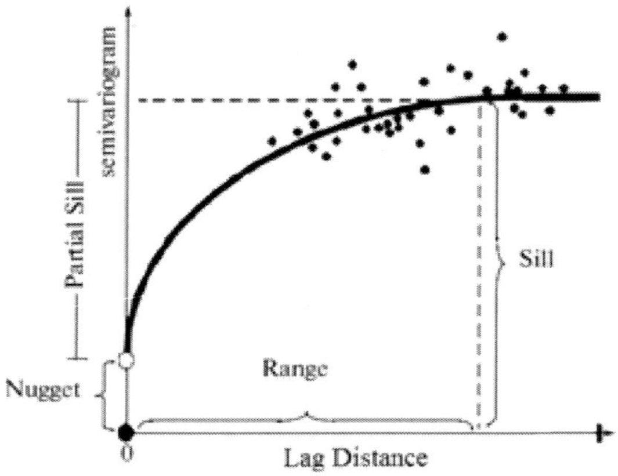

Figure 3. The semivariogram model and its parameters.

One of the semivariogram parameters as seen in Figure 3, nugget, is the point where the theoretical curve intercepts vertical axis at h=0 point. It does not affect the prediction value, it only causes change in the Kriging variance. Partial Sill is the vertical scale value for the structural components of the variogram. Range is the horizontal distance of the variogram. After this length, the data are uncorrelated with each other. The variogram or covariance values remain stable after this length. The change in the range causes also the change of interpolation value. The lLag distance is the horizontal distance between the sampling points and inter sampling points or interpolation points. The sill is the total vertical scale value of the variogram function. It does not affect the prediction value, it only causes change in the Kriging variance (Akyurek et al., 2013).

The Kriging, which is the geostatistical analysis method at Geostatistics, is used to determine the quadrennial air quality int the area examined, and it is carried out by aiming to map spatial distribution of the pollutant parameters by the geostatistical analysis.

Results and Discussion

The annual mean values that are calculated from the pollutant parameters (PM_{10}, SO_2 and NO_2) and obtained as hourly average from the

45 Air Quality Monitoring Centers in the Marmara Region between the years of 2014 and2017, are used in the study. Firstly, the comparison of the limit values is carried out with the calculated annual means. Accordingly, the number of the stations that exceed the limit values according to the Turkish Limit Values are detected and shown in the Table 2.

When the number of the stations, which exceed the limit values are examined, for the particular matters and SO_2, it is found that they are the air quality stations in which industrial plants are located generally. It is observed that the limit values exceeded especially in the stations that were in Istanbul, Bursa, Tekirdag and Kocaeli. Also, the PM_{10} and SO_2 values were calculated far above the limit values in Kesan province that used solid fuel for domestic heating and the Can province that had thermal power plant within its borders, during the time period when the data were collected. It is observed that the SO_2 values that were calculated in Kesan and Can were ten times higher than the limit values. When the stations that exceeded the NO_2 limit value were examined, it was evident that these stations were the stations in the provinces that generally had high values in terms of population. The main sources of the NO_2 were the fuels that were used by the vehicles which re used for transportation and industrial plants. When the quadrennial period was examined, the NO_2 value was calculated above the limit values especially in the stations in the provinces that weres crowded both in terms of population and industrial plants such as Istanbul and Bursa.

Table 2. The number of stations that exceed the Turkish Limit Values by years

	PM_{10}	SO_2	NO_2
2014	18	6	8
2015	20	8	7
2016	18	13	8
2017	21	10	8

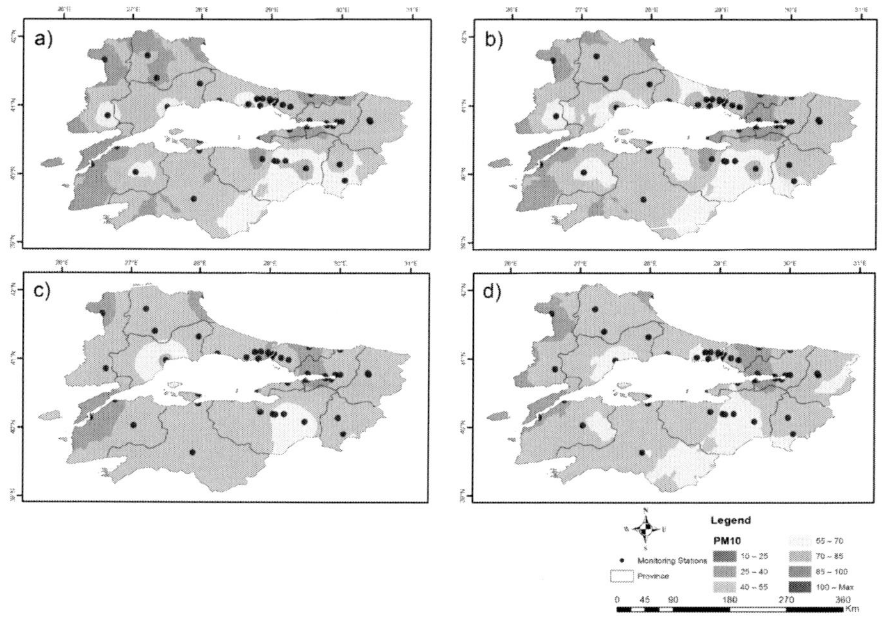

Figure 4. The PM_{10} Pollutant's Distribution Maps; a) 2014, b) 2015, c) 2016 and d) 2017.

In the creation of the spatial distribution maps, the weights and possibilities that were determined by the semivariograms by the help of ordinary Kriging method, were used. With the help of this method, the pollution distribution maps of three pollutant parameters were created belonging to four different years (from 2014 to 2017).

The distribution maps that belonged to the PM_{10} pollutant parameter are shown in Figure 4. In Figure 4-a, in the distribution map that belonged to 2014, it is seen that Bursa province, which is one of the industrial cities of the Marmara Region, and Istanbul Esenyurt district were high in terms of the PMs . Also, it was observed that PMdensities in the air were above the limit values in Kesan, Can and Bozuyuk districts and also in Tekirdag city center. In Figure 4-b, it the distribution map that belonged to 2015, it is seen that PM density in Esenyurt and Tekirdag city center was higher than the previous year. In other centers, on the other hand, it is observed that it remained at similar levels. In Figure 4-c, in the distribution map belonging to 2016, it is seen that pollution level went down the limit level and below in the other stations except in Tekirdag city center and the Bursa province. In Figure 4-d, in the distribution map that belonged to 2017, it is seen that pollution level maintained the tendency of the previous year.

The biggest source of the SO_2 pollutant is the solid fuels that are used for domestic heating. Because of the fuels that have high Sulphur rate are not burned properly, the Sulphur amount extracted to the atmosphere is measured very high in the locations that use solid fuels for domestic heating. The SO_2 pollutant parameter's distribution maps are shown in Figure 5. In Figure 5-a, the distribution map that belonged to 2014 is shown. It is seen that the values were exceedingly above the limit value in Edirne's Kesan district that used stove and its derivatives for domestic heating during the time of data collection and Canakkale's Can district that had coal-fired thermal power plant, which was used for power generation within its borders. Also, it is found out that calculated values were above the limit values again in the Bursa province and Kocaeli's Kandıra district. The distribution maps that belonged to 2015 and 2016 are shown in Figure 5-b and c. It is seen that the values in Kesan and Can districts were still far above the limit values although the other stations went down the limit value level. In Figure 5-d, it is observed that the SO_2 level measured in the stations, fell down thanks to the precautions taken. It is seen that while the measured pollutant values reached the limit values in Can, they were at lower levels compared to the previous years in Kesan.

The biggest sources of the NO_2 pollutant are the fuels that are used at the transportation vehicles, industrial plants and fluid fuels that are used for the domestic heating. The distribution maps that belong to the NO_2 pollutant parameter are shown in Figure 6. In Figure 6-a, in the distribution map that belonged to 2014, it is seen that the NO_2 pollutant parameter was close to the limit value and above in Istanbul, Bursa, Kocaeli and Tekirdag provinces that had industrial plants intensively, were the highly populated provinces and had the highest migration rates every year. In Figure 6-b and c, a similar situation is seen in the distribution maps that belonged to 2015 and 2016. But in Figure 6-d, in the distribution map that belonged to 2017, it is seen that the measured values distributed to the entire provincial border with a rapid increase especially for Istanbul and Bursa.

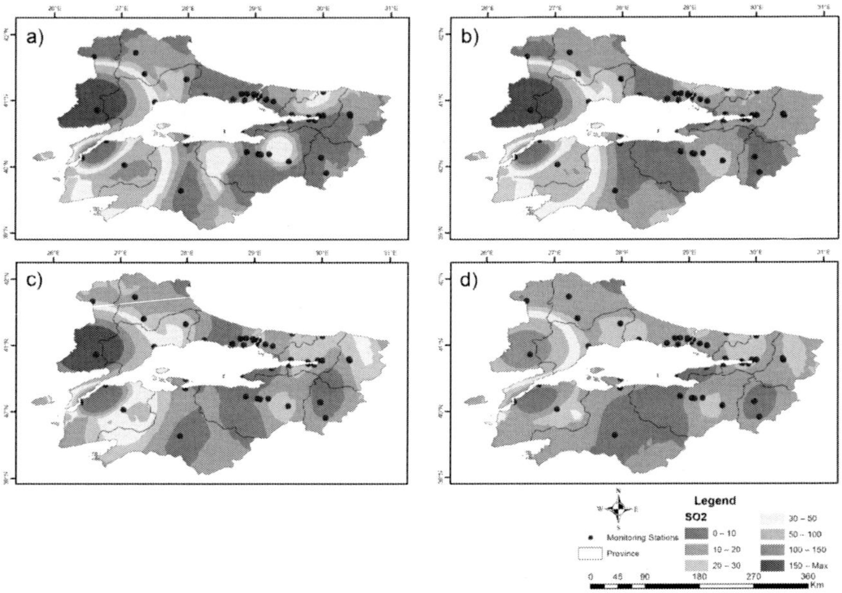

Figure 5. The SO_2 Pollutant's Distribution Maps; a) 2014, b) 2015, c) 2016 and d) 2017.

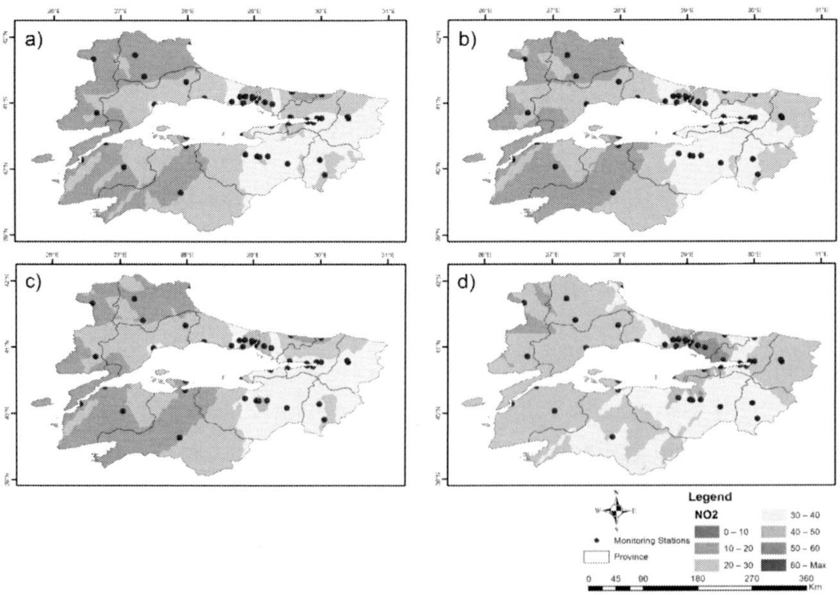

Figure 6. The NO_2 Pollutant's Distribution Maps; a) 2014, b) 2015, c) 2016 and d) 2017.

CONCLUSION

In this study, the pollutant parameters' distribution maps by years were created by using spatial analysis of the air pollutant parameters that were measured by 45 Air Quality Monitoring Stations in the Marmara Region between the years of 2014 and 2017. With Geostatistics technique, the spatial pollution distribution maps were created by the identification of spatial patterns regarding to the air quality parameters and the station-based statistics were evaluated with the dynamic mapping. The findings of this study revealed that the Geostatistics-based spatial statistical applications are an effective tool for the air quality management.

It was observed in the maps, which were used in this study that PMs showed higher distributions in the locations that had more amount of industrial plants and they reached to high values also in the locations that used solid fuels for domestic heating. When the SO_2's distribution maps were examined, the Kesan and Can districts reached to excessive values in terms of this pollutant but it was observed in the distribution maps that precautions taken were effective. When the NO_2's distribution maps were examined, it was observed that the cities that had higher population density, had higher values in terms of this pollutant. The effect of population on this parameter might be clarified with intensive vehicle traffic, fluid fuels that were used for heating and intensive industrial plants.

Air is a vital environmental element. This environmental element, which is vital to maintain life, is essential for human life as a physical factor. The air is breathed between ten thousand and twenty thousand liters per person in a day. The air that is used in such a high rate must have high quality. Low air quality causes fatal effects. Air quality needs to be examined regularly and kept under control. Geostatistics is an effective tool for monitoring and examining air quality. The pollution distribution maps that are created with the help of Geostatistics, turns into a source of information that provides prominent information for air quality determination and enhancement. It provides useful information to the decision-makers for the precautions that will be taken in the future. With the help of spatial mapping that can be updated instantly, people will be provided with information about the quality of the air they breathe and decision makers can provide a healthier environment and clean air for the people who live in these places.

REFERENCES

Akyürek, Ö., Arslan, O., Karademir A. (2013). SO_2 ve PM_{10} Hava Kirliliği Parametrelerinin CBS ile Konumsal Analizi: Kocaeli Örneği, *TMMOB Coğrafi Bilgi Sistemleri Kongresi*, 11-13 Kasım, Ankara, Türkiye. [Locational Analysis of GIS Parameters of SO_2 ve PM_{10} Air Pollution, *TMMOB Geography Information Systems Congress*, 11-13 November, Ankara, Turkey].

Arslan, O., Akyürek, Ö. (2018). Spatial Modelling of Air Pollution from PM_{10} and SO_2 concentrations during Winter Season in Marmara Region (2013-2014), *International Journal of Environment and Geoinformatics*, 5(1), 1-15.

Atkinson, P. M. (2002). Spatial Statistics, Editors: Stein A., Van Der Meer F., Gorte B., Spatial Statistics for Remote Sensing, 1 ed., *Kluwer Academic*, Dordrecht, 57–81.

Briggs, D., Beale, L., de Hoogh, C., Gulliver, J. and Mitchell, J. F. (2003). Beyond the buffers: advancing the use of GIS to model exposures to environmental pollution: Isee-478. *Epidemiology*, 14(5), 95.

Çetin, Ş. (2006) *Kocaeli ilinde NOx Emisyon Dağılımlarının Modellenmesi,* Doktora Tezi, Fen Bilimleri Enstitüsü, Kocaeli Üniversitesi, Türkiye. [*The Modelling of NOx Emission Distribution in the City of Kocaeli*, Unpublished Doctoral Dissertation, Institute of Applied Sciences, Kocaeli University, Turkey].

Chen, W., Yan, L. and Zhao, H. (2015). Seasonal Variations of Atmospheric Pollution and Air Quality in Beijing. *Atmosphere*, 6, 1753-1770.

David, J. B., Collins, S., Elliott, P., Fischer, P., Kingham, S., Lebret, E., Pryl, K., Van Reeuwijk, H., Smallbone, K. and Andre, V. (1997). Mapping urban air pollution using GIS: a regression-based approach.

International Journal of Geographical Information Science, 11, 699-718.

Demiraslan, K. O., Akıncı, H. (2018) CBS ve Hava Kalitesi Verileri kullanılarak Marmara Bölgesinin Kış Sezonunda Hava Kalitesinin Değerlendirilmesi, *Doğal Afetler ve Çevre Dergisi*, *4*(1), 11-27. [An Evaluation of Marmara Region Winter Season Air Quality via GIS and Air Quality Data, *Natural Disasters and Environment Journal*, *4*(1), 11-27].

Elbir, T. (2004). A GIS based decision support system for estimation, visualization and analysis of air pollution for large Turkish cities. *Atmospheric Environment*, 38, 4509-4517.

Esperanza, P., Bencît, F., Hervé, P. (2004). Using Geostatistics to Assess the Area of Spatial Representativity of Air Quality Monitoring Stations. In: Sanchez-Vila X., Carrera J., Gómez-Hernández J. J. (eds) geoENV IV — Geostatistics for environmental applications. *Quantitative Geology and Geostatistics*, vol 13. Springer, Dordrecht.

Fedra, K. & Haurie, A. (1999). A decision support system for air quality management combining GIS and optimization techniques. *International Journal of Environment Pollution*, 12, 125-146.

Hee-Jae, K. & Myung-Jin, J. (2014). Analysis on relationship between urban development characteristics and air pollution level – a case of seoul metropolitan region. *Korea Planners' Association*, 49(7), 151-167.

Kim, J., Ghim, Y. S., Han, J. S., Park, S. M., Shin, H. J., Lee, S. B., Kim, J., Lee, G. (2018). Long-term trend analysis of korean air quality and its implication to current air quality policy on ozone and PM_{10}, *Journal of Korean Society for Atmospheric Environment*, 34(1), 1-15.

Lin, M. D. & Lin, Y. C. (2002). The application of GIS to air quality analysis in Taichung city, Taiwan, ROC. *Environmental Modelling & Software*, 17, 11-19.

Matejicek, L. (2005). Spatial modeling of air pollution in urban areas with gis: a case study on integrated database development. *Advances in Geosciences*, 4, 63-68.

Moral, F. J., Valiente, P. F., Lopez, F. (2010). Geostatistical analysis and mapping of ground level Ozone in medium sized urban area. *International Scholarly and Scientific Research & Innovation*, 4(1), 46-57.

Partigöç, N. S., Çubukçu, K. M. (2017). Hava Kirliliği ve Kent İlişkisine Ampirik Bakış: Ekolojik Sürdürülebilirlik Ekseninde Bir Değerlendirme, *Akademia Disiplinlerarası Bilimsel Araştırmalar Dergisi, 3*(2), 28-45. [An Emprical Perspective to Air Polution and City Relationship, *Academia Interdisciplinary Scientific Research Journal, 3*(2), 28-45].

Rohde, R. A. and Muller, R. A. (2015). Air Pollution in China: Mapping of Concentrations and Sources. *PLOS ONE, 10*(8).

Song, S. J. (2008). A GIS Based Approach to Spatio-Temporal Analysis of Urban Air Quality in Chengdu Plain. *The International Achieves of the Photogrammetry, Remote Sensing and Spatial Information Sciences*, 37(B7), Beijing.

Xiao, F., Qi, L., Yajie, Z., Jingje, W., Heming, L., Ruofeng, X. (2014). Formation and Dominant Factors of Haze Pollution Over Beijing and Its Peripheral Areas in Winter. *Atmospheric Pollution Research*, 5(3), 528-538.

Yeşilkanat, C. M., Kobya, Y., Taşkın, H., Çevik, U. (2014). Jeoistatistik Tahmin ve Simulasyon Yöntemleri ile Artvin İlindeki Doğal Kaynak Suları için Toplam Alfa ve Toplam Betanın Ara Değer Modellemesi ve Haritalanması, Cumhuriyet Üniversitesi Fen Fakültesi Fen Bilimleri Dergisi, 35(4), 11-35. [Interpolated Modelling and Mapping of Gross Alpha and Gross Beta for Natural Spring Water in Artvin Province by Using Geostatistical Prediction and Simulation Methods, *Cumhuriyet University Faculty of Science Science Journal (CSJ), 35*(4), 11-35].

In: Advances in Health and Natural Sciences ISBN: 978-1-53614-639-4
Editors: Burcu Yuksel et al.

Chapter 2

OBESITY, OBESITY SURGERY AND NURSING CARE

Züleyha Şimşek Yaban
Nursing, Kocaeli University Faculty of Health Sciences, Turkey

ABSTRACT

Obesity is not just an emerging health problem, it also leads to other public health problems, including: diabetes, hypertension, cardiovascular diseases, cancer, depression, infertility, and it negatively affects social life as well. In the simplest terms, obesity is an energy irregularity characterized by abnormal or excessive fat accumulation.

The increase in obesity incidence is a common situation for all countries that is parallel with their development level and increased consumption of ready-to-eat foods. According to the World Health Organization (WHO), obesity has increased nearly threefold worldwide since 1975. In 2016, more than 1.9 billion adults over 18 years of age were overweight, and over 650 million of them were obese. The obesity rate in Turkey was 15.2% in 2008, and it increased by 31.1% and reached 19.9% in 2014. Globally, many factors contribute to increasing in obesity rates due to reduced physical activity as a result of jobs that require minimal physical activity become more common, less mobility, changing modes of transport, the consumption of fatty and high calorie foods and increased urbanization.

Obesity is an important public health problem in our country and worldwide, and it causes a variety of physiological, psychological and social complications by negatively affecting obese people's quality of life and many organ systems in their bodies. Although diet, physical activity, psychological support and pharmacotherapy are commonly used treatments for obesity, obesity surgery (bariatric surgery) has recently emerged as a contemporary treatment modality.

The success of obesity surgery requires a multidisciplinary team approach including surgeons, psychiatrists, nutritionists, nurses, physiotherapists and other health professionals from the moment patients apply to hospital until being discharged. It is also very important to meet special needs due to the accompanying diseases of obesity surgery patients.

Providing personalized, individualized and holistic nursing care and counseling services to the patients who will undergo obesity surgery patients to reduce possible risks in their care, and preventing complications and difficulties before, during and after surgery are very crucial. These services are also important to increase patients' quality of life.

Keywords: obesity, obesity surgery, bariatric surgery, nursing care

INTRODUCTION

Nowadays, obesity is considered as a chronic, progressive and life threatening disease because of the comorbidities and obesity-related psychological problems. Obesity which is defined as "excessive fat accumulation in the body to the extent that derange health" by the World Health Organization (WHO), is a disorder that needs to be treated endocrinally, metabolically, and psychologically. Normally, fat tissue constitutes 15-20% of body weight on average in adult males while it constitutes 25-30% of body weight on average in adult females. When the ratio of fat tissue exceed 25% of body weight in males and 30% of body weight in females, this condition is called as obesity (Bozbora, 2002; Turkish Public Health Institution, 2013).

The prevalence of obesity which has become an increasingly important health problem, increases in all ages and socio-economic groups in all developed and developing countries. According to the obesity rates in the Organisation for Economic Cooperation and Development (OECD) countries, the ratio of obese persons in the adult population is 19.5%. This rate varies according to countries, it is lower than 6% in Korea and Japan and higher than 30% in Hungary, United States of America, New Zealand and Mexico. More than a quarter of adult population is affected by obesity in the United Kingdom, Chile, Australia, Canada and South Africa. The ratio of overweight and obese individuals has increased in the United States, France, Mexico, Canada and Switzerland while it has not changed in Korea, Spain, England and Italy within last decade (OECD, 2017). In Turkey, the obesity rate was 15.2% in 2008, increased by 31.1% and reached to 19.9% in 2014 (TUIK, 2015). According to the OECD 2017 data, the prevalence of obesity in Turkey (22.3%) is higher compared the average obesity prevalence in the OECD countries. Obesity rates are expected to be high especially in the United States (47%), Mexico (39%) and England (35%) in the year of 2030 (OECD, 2017).

Obesity is accepted as a natural consequence of modern life by many individuals while it is usually related with diseases such as cardiovascular diseases, metabolic syndrome, type 2 diabetes mellitus (DM), a decrease in physical activity due to lifestyle especially in transportation, production and agriculture along with advancing technology, the changing eating habits in modern life play roles in the increasing prevalence of obesity. Fast-food is a diet being rich in terms of fast-eating unhealthy foods, carbohydrates and refined sugar and which is poor in terms of vegetable fiber, leads to obesity. In addition, the widespread use of high-tech devices (mobile phones, televisions, computers, home cinema systems, etc.) that easily fill our leisure time, has an important contribution to the increase of obesity. In addition, endocrine disorders such as diabetes mellitus, hypothyroidism, growth hormone deficiency and drugs that disrupt metabolic control such as antipsychotics, glucocorticoids, anabolic steroids also play an important role in increasing prevalence of obesity (Turkey

Endocrinology and Metabolism Society, 2017; Turkish Public Health Institution, 2013).

Nowadays, there are three methods to determine obesity. These three methods are ideal weight, body mass index (BMI or Quetelet index) and the measurement of subcutaneous fat. Individuals having a body weight which is 10% higher than ideal weight, are defined as overweight while individuals having a body weight which is 20% higher than ideal weight, are defined as obese. BMI is a practical method used to classify obese people and to plan treatment methods (Asoglu, 2008; Bozbora, 2002). The World Health Organization (WHO) defines overweight and obesity based on body mass index (BMI = Weight [kg] / Height [m^2]). According to this definition, individuals having a BMI between 18.5-24.99 kg/m^2 are classified as normal weight; individuals having a BMI between 25.0-29.9 kg/m^2 are classified as overweight; individuals having BMI $\geq$ 30 kg/m^2 are classified as obese; individuals having a BMI between 30-34.9 kg/m^2 are classified as class 1 obese; individuals having a BMI between 35-39.9 kg/m^2 are classified as class 2 obese; individuals having a BMI $\geq$ 40 kg/m^2 are classified as class 3 obese (Turkey Endocrinology and Metabolism Society, 2017; WHO 2018). In the measurement of subcutaneous fat tissue, which is another method for the diagnosis of obesity, the sum of the thickness of skin across the scapula and triceps which is >55 mm in females and >32 mm in males is defined as obesity (Bozbora, 2002).

Obesity creates a fertile environment for the development of many health problems such as cardiovascular diseases, hypertension, diabetes, certain types of cancer, respiratory system diseases, musculoskeletal system diseases. It also increases health care expenditures and negatively affects the quality and duration of life (Turkey Endocrinology and Metabolism Society, 2017). Advanced obesity is often called as morbid obesity because of the diseases associated with obesity. Morbid obesity is a metabolic disease and is associated with numerous medical problems. In fact, most of these problems are not observed in the absence of obesity while many health problems arise with weight gain. The most common problems are arthritis and degenerative joint diseases; at least 50% of the patients who undergo surgery due to morbid obesity have these accompanying diseases. Premature deaths are more frequent in children of obese individuals; the death rate of morbid obese individiuals is 12 times higher compared to normal-weight individuals in 25-35 years old male individuals. Obesity is observed in 25% of asthmatic patients, 30% of hypertensive patients, 20% of diabetic patients and 20-30% of the patients with gastroesophageal reflux. The incidence of these comorbid diseases increases according to the duration of obesity and age. The number of comorbidities (Bozbora, 2002; Sakrak and Dikmen, 2010), socioeconomic problems, psychological disorders and premature death risk increase in patients over 50 years of age (Asoglu, 2008).

TREATMENT OF OBESITY

The goal of obesity treatment is to alleviate obesity-related morbidity and mortality risks, to bring individuals adequate and balanced nutrition habits and to improve the quality of life (Turkey Endocrinology and Metabolism Society, 2017). Obesity is a global pandemic and is also considered as a treatable and preventable health problem. It can be treated by medical nutrition (diet), physical exercise, behavioral change, pharmacological and surgical methods (Alican, 2007). The diet succeeds only in a small part of the patients and does not take a long time for the patients to regain their lost weight. An effective and safe pharmacological agent has not been developed in the treatment of obesity. Pharmacological preparations in use are not effective or not preferred due to their adverse effects. In this context, it has been stated that diet, physical exercise and behavioral change are indispensable in obese patients while pharmacological and surgical treatment can be applied if required. Surgical treatment may be effective for the treatment of morbid obese patients (Asoglu, 2008; Yuksel, 2016).

Medical Treatment

Obesity treatment is a combination therapy that includes lifelong adequate and balanced nutrition, increased physical activity and behavioral therapies. Essentially, medication is not recommended for the treatment of obesity at the beginning. However, drug treatment may be applied temporarily for selected patients (patients in the high-risk group (BMI ≥ 30 kg/m^2)) by an endocrinologist (Turkish Public Health Institution, 2013).

In obesity treatment, behavioral change therapy is absolutely necessary along with nutritional therapy and increased physical activity. In combined treatment, nutritional therapy, exercise and behavioral change therapy are used in combination and provide great success in both weight loss and maintenance of weight loss. The goals of medical nutrition therapy in obesity are:

1. To provide adequate and balanced nutrition which is appropriate for the individual's age, gender, physical activity status, life style and physiological condition.
2. To provide proper and lasting nutrition habits instead of wrong nutrition habits.
3. To decrease the body weight to the desired level (this level may be an ideal weight or a weight over ideal weight).
4. To stop regaining weight, thus individuals lost body weight and to keep it at the desired level lastingly.
5. To maintain normal growth and development in children (Turkey Endocrinology and Metabolism Society, 2017).

Surgical Treatment

The surgical treatment of obesity is the only proven method for achieving long-lasting weight loss by controlling and potentially resolving life-threatening comorbidities and improving the quality of life of morbid obese patients at the same time. It was showed that the expected length of life increased by 12 years after the surgical treatment of obesity (Frantzides et al., 2013). Obesity surgery (bariatric surgery) is one of the increasingly popular obesity treatment methods in the world in the last decade. The number of patients undergoing bariatric surgery increase day by day because obese patients lose weight rapidly in addition to a decrease in comorbid diseases after bariatric surgery (Yuksel, 2016).

The aim of surgical treatment is to reduce morbidity and mortality due to obesity and to improve metabolism and organ functions. Thus, obesity-related comorbidity, drug expenditure and time lost due to being ill decrease while the quality of life improves (Turkey Endocrinology and Metabolism Society, 2017).

Despite the significant developments in technique, care and follow-up in bariatric surgery, the risk and perioperative complications have not been alleviated. Therefore the selection of appropriate patient and appropriate technique is very important (Turkey Endocrinology and Metabolism Society, 2017). A bariatric surgical team consists of an endocrinologist, nutritionalist, anesthetist, operating room nurse, operating room technician, nurse care coordinator (advanced nurse or a well-educated nurse), a psychiatrist, a psychologist, cardiac, pulmonary, gastrointestinal, endocrine, musculoskeletal and neurological specialists (Frantzides et al., 2013; Sakrak and Dikmen, 2010; Turkish Public Health Institution, 2013; Yurdakul, 2015).

The selection of an appropriate patient should be meticulously performed according to the recently adopted criteria of the National Health Institution (NIH) independently of the type of surgery, to be able to perform bariatric surgery. For performing bariatric surgery, a patient must have a Body Mass Index (BMI) ≥ 40 kg/m^2; if a patient's BMI is >35 kg/m^2, he/she must have obesity-related comorbidities (Type 2 diabetes, severe HT, arthritis, sleep apnea syndrome, etc.), inability to control it with medical treatment and lifestyle changes, previous unsuccessful attempts with non-surgical methods (diet, medicine, etc.), an age between 18 and 60 years and the duration of obesity >5years (Bozbora, 2002; Frantzides et al., 2013; Sakrak and Dikmen, 2010; Turkey Endocrinology and Metabolism Society, 2017). The inability to adapt daily life conditions because of obesity and having some psychological problems due to obesity are relative indications for surgical intervention. Inability to follow the postoperative follow-up protocol, excessive amounts of medication, alcohol or drug use, mental retardation, the presence of brain tumors affecting the satiety center and severe psychiatric disease (major depression and psychosis) or advanced cardiac disease are among the contraindications for surgical intervention (Bozbora, 2002; Turkey Endocrinology and Metabolism Society, 2017).

The surgical treatment of obesity (bariatric surgery) is preferred more compared to the past in which applied surgical procedures lead to weight loss by causing malabsorption or restricting food intake. There are also combined surgical procedures that use both methods together (Turkey Endocrinology and Metabolism Society, 2017).

a) Restriction of Food Intake: The food storage capacity of the stomach is reduced by resection, bypass or narrowing the upper part of the stomach of the patient, thus the patient consumes less food. Vertical band gastroplasty (VBG), sleeve gastrectomy (SG) and laparoscopic adjustable gastric band (LAGB) are restrictive surgical procedures. There is no change in the absorption function of small intestine after these procedures (Turkey Endocrinology and Metabolism Society, 2017).
b) Surgery for malabsorption: Nutrient absorption is reduced by shortening the length of the small intestine by bypassing the absorbing area or by distracting the bileopancreatic secretions facilitating absorption. Jejunoileal bypass (JIB) and duodenal switch (DS) are these type operations. However, protein-calorie malnutrition and the deficiency of some micronutrients develop when the malabsorption is too high (Turkey Endocrinology and Metabolism Society, 2017).
c) Restrictive and malabsorptive, combined effective procedures: Roux-en-Y gastric bypass (RYGB), biliopancreatic diversion (BPD) and bilopancreatic diversion / duodenal switch (BPD / DS) are these type operations. In RYGB, malabsorption is provided

with small intestine reconstruction while restriction is provided with small stomach pouch (Turkey Endocrinology and Metabolism Society, 2017).

Totally, 468,609 bariatric procedures was performed in 2013 while 95.7% of the procedures were carried out laparoscopically according to the International Federation for the Surgery of Obesity and Metabolic Disorders (IFSO) consisting of 50 nations or national groups. The highest number (n = 154, 276) of procedures was performed in the USA/Canada region. Roux-en-Y gastric bypass (RYGB) was the most common procedure in the world with the ratio of 45%; sleeve gastrectomy (SG) was the second most common procedure with the ratio of 37%; adjustable gastric banding (AGB) was the third one with the ratio of 10%. The prevalence of SG increased from 0% to 37% worldwide from 2003 to 2013 while the prevalence of AGB decreased from its' peak to 68% from 2008 to 2013. SG is currently the most frequently performed procedure in the USA/Canada and in Asia/Pacific regions and second most frequently performed procedure after RYGB in the Europe and Latin America/South America regions (Angrisani et al., 2015).

NURSING CARE IN OBESITY SURGERY

The presence of comorbidities such as diabetes, hypertension and respiratory problems caused by obesity requires that the care of the patients undergoing obesity surgery before, during and after the surgery should differ from the care of other patient groups.

Preoperative Nursing Care

- It is very important that the patient is prepared well before the surgery for success in the obesity surgery. The patient's anxiety must be reduced by giving information about the needs of the patient before, during and after the surgery (Bozbora, 2002).

- Other risk factors leading to obesity such as gender, age, educational level, occupation, marital status, monthly income level, number of pregnancies and deliveries, smoking, alcohol use, psychosocial factors and chronic stress, drugs (antidepressants, antiepileptics, corticosteroids, progestational steroids, hormonal contraceptives, etc.) and the symptoms of the diseases leading to secondary obesity, obesity complications, the symptoms of the comorbidities (type 2 diabetes, HT, dyslipidemia, cardiovascular and respiratory diseases, joint diseases, non-alcoholic fatty liver disease (NAFLD) and sleep disorders) must be examined (Turkey Endocrinology and Metabolism Society, 2017).
- During the preparation period, blood tests (hemoglobulin, hematocrit and whole blood count, prothrombin (PT) and partial thromboplastin time (PTT), cholesterol, triglyceride, urea, creatinine, electrolyte levels), urinalysis, radiological examinations (chest X-ray, gallblader and liver ultrasonography), cardiogram and echography and tests for special conditions due to obesity must be performed (Bozbora, 2002; Garza, 2003; Sakrak and Dikmen, 2010).
- Determining the level of nutrition during the routine examination of the patient is very important in terms of the effectiveness of the treatment methods to be performed. This assessment includes the size and weight of the patient, nutritional anamnesis, anthropometric measurements and physical examination (Bozbora, 2002; Neil, 2013; TUIK, 2015). The food preferences of the patient, the amount and time of eating must be found out. The status of having a chronic, systemic disease is important in evaluating the nutritional status of the patient (Bozbora, 2002).
- In the morbid obese, important physiopathologic changes are observed in the gastrointestinal cardiovascular and respiratory systems. In obese patients, oxygen consumption and carbon dioxide production increase in the respiratory system because of

the high energy consumption and metabolic demand (Tanaka and Peniche 2009). When sleep apnea is diagnosed, a treatment such as continuous positive airway pressure (CPAP) can be started before surgery and some arrangements are made to monitor O_2 saturation of patients via pulse oximetry; CPAP can be continued after surgery (Garza, 2003).

- In the preparation of the patient for laparoscopic bariatric surgery, mechanical bowel cleaning is performed for 2 days. Patients are free to drink water until one day before surgery therefore they are protected from dehydration (Sakrak and Dikmen, 2010).
- The patient must take a shower to clean the skin before the obesity surgery. Skin hairs must also be removed with epilators or hair removal creams if necessary.
- Skin risk assessment (e.g., Braden Score) must be performed at hospital admission, with continuous monitoring during hospitalization and follow-up care of infection, ulceration of skin folds, irritation, atypical and pressure ulcers (Mulligan et al., 2009).
- Chairs, operating room tables, stretchers, wheelchairs, commodes, patient scrubs and sphygmomanometer cuffs must be prepared in sizes which are suitable for the patient before surgery. In order to ensure the confidentiality of the patient, the patient must be allowed to live alone in a private room (Asoglu, 2008).
- Before the surgery, wide vein path are opened with a Foley catheter for each patient. In super obese patients, Foley catheter placement maybe difficult due to covering of the perineum by the belly. In obese patients with increased subcutaneous fat tissue, central venous catheterization should be preferred because of the difficulty of finding a peripheral vascular access and the difficulty of maintaining continuity in the postoperative period. Arterial catheterization for invasive blood pressure monitoring facilitates blood pressure monitoring in the early postoperative period in intensive care (Asoglu, 2008; Otmalaz and Ece, 2016).

- The importance of early post-operative ambulance, deep breathing and coughing exercises and the use of spirometers must be emphasized and taught to the patient (Garza, 2003).
- Prophylaxis is provided against deep vein thrombosis (DVT) and pulmonary embolism. For this purpose, important measures are taken such as early mobilization within 4-6 hours after surgery, the use of pressure socks and shoes to pressure lower extremities and subcutaneous administration of low molecular weight heparin. High-risk patients need to administer heparin for 2 weeks after discharge therefore subcutaneous administration must be taught to high risk patients before discharge (Sakrak and Dikmen, 2010).
- Patients' post-operative anticipations must be discussed with them, to ensure that their goals are realistic. They must be informed about all possible complications in detail and their written consents should be obtained (Turkey Endocrinology and Metabolism Society, 2017).
- The preoperative nurse needs to note abnormalities about the patient, adds them in the care plan and shares their concerns with the surgeons and anaesthesiologists.

Intraoperative Nursing Care

- While the standard operating tables have a weight limit of 200 kilograms, the weight limit of the tables designed for bariatric surgery is 450 kilograms. Since most of the bariatric interventions may be required to position the patient, it must be ensured that the patient does not fall from the table and is safe (Asoglu, 2008). The nurse must ensure that adequate number of personnel are in the operating room to safely move and position the patient (Neil, 2013).
- The healthcare staff must be attentive to protect the patient from pressure ulcers and nerve damages by supporting pressure points with pads. This is especially important for advanced obese and

diabetic patients (Asoglu, 2008). In the intraoperative period, the nurse, who is primarily responsible for the safety and care of the patient, should take necessary precautions and applied interventions to prevent pressure sores during the operation.

- The perianesthesia nurse needs to be prepared to help the positioning of the patient for ideal tracheal intubation (Mulligan et al., 2009).

Postoperative Nursing Care

- Vital findings that provide a potential sign to important problems should be evaluated carefully. Patients should be monitored for the first 24 hours after surgery for the complications such as hypoxia, hypercapnia and sleep apnea (Turkey Endocrinology and Metabolism Society, 2017).
- Oxygen saturation should be monitored by pulse oximeter for the first 24-48 h after surgery. So emergency tracheostomy may be required, the necessary equipment have to be available at the bedside.
- The pain management is very important for the patients' early ambulation, spirometer use and participation to deep breathing and coughing exercises. A PCA (patient controlled analgesia) device must be provided to the patient, thus the patient keep his/her pain under control. Pain control is important for ambulation and breathing exercises, both of them are important to protect the patient from respiratory problems such as atelectasis or infection and thrombosis (Mulligan et al., 2009; Owers et al., 2012; Sakrak and Dikmen, 2010).
- Morbid obese patients have decreased blood supply due to excessive fat tissue while wound healing is also delayed in these patients. Therefore nurses must perform daily wound care and dressing to prevent delay in wound healing due to infection, and

observe wounds in terms of color and drainage (Barth and Jenson, 2006).

- The prevalence of venous thromboembolism is higher in obese patients, because of that ambulation and physical activity are important. Nurses must observe the patient for venous thromboembolism symptoms (e.g., swelling, pain or tenderness, warmth, or redness in legs or one leg) and note information about their observations in the postoperative teaching materials. Compression socks, early and frequent ambulation, leg exercises in the bed once in two hours, anticoagulant or low-molecular-weight heparin must be applied as a common prophylactic intervention to prevent DVT (Barth and Jenson, 2006; Mulligan et al., 2009; Neil, 2013; Sakrak and Dikmen, 2010; Tanaka and Peniche 2009).
- It is very important to monitor fluid intake and fluid loss of the patient and proper liquid replacement system functions (Barth and Jenson, 2006; Mulligan et al., 2009). Fluid electrolyte imbalances due to overfeeding and heavy diets should not be overlooked.
- In the postoperative period, cardiac rhythm and blood pressure must be monitored at regular intervals to be aware of the increase in heart rhythm, one of the symptoms of complications such as hemorrhage and anastomosis (Barth and Jenson, 2006; Neil, 2013).
- Diabetic subjects must be monitored in terms of blood glucose level and hypoglycemia (Turkey Endocrinology and Metabolism Society, 2017).
- It is important to position the patient. The weight of the abdominal contents compress the diaphragm and makes breathing more difficult, therefore the standard supine and Trendelenburg positions must be avoided. For preventing respiratory distress, the patient's head and neck must be raised 30°-45° to be flat, and the imporovement of tidal volume should be allowed (Barth and Jenson, 2006; Mulligan et al., 2009; Neil, 2013).
- In patients undergoing obesity surgery, sleep apnea incidence is high. Nurses must be aware of potential undiagnosed sleep apnea in patients undergoing obesity surgery (Neil, 2013).

- The most lethal complication of bariatric surgery is leakage from the gastrointestinal system. Tachycardia, occasional tachypnea and agitation can be the only signs of a serious intraabdominal problem (Sakrak and Dikmen, 2010).
- The ambulation of the patient must be started as early as possible, the nurse must ensure that required equipments are available to help the patient for upstanding and position changing. Use of special bariatric beds, air-assisted transfer devices, mechanical patient lifts, bed trapezes and other equipment helps ensuring the safety of nurses and patients in the mobilisation and positioning of patients (Barth and Jenson, 2006; Mulligan et al., 2009; Neil, 2013).
- In restrictive and disruptive surgery or combined techniques, oral feeding is not started at early period due to impaired gastrosintestinal integrity and the adaptation of patients to the newly created gastrointestinal system anatomy takes time. Nurses' support for nutrition during the hospitalisation period reduces patients' anxiety (Otmalaz and Ece, 2016).
- The written materials which contains information about the progression of the diet, appropriate food choices, bowel habits changes, drugs, physical activity, wound care, signs and symptoms to report to the physician, follow-up visits, vitamin and protein supplements, and the importance of drinking required amount of liquid (hydration) after discharge, should be provided to patients (Garza, 2003).
- Long term psychological support should be provided to the patients from the preoperative period to postoperative period, so the problems such as low self esteem, decreased socialization, intensive shame and fear can be alleviated (Barth and Jenson, 2006).
- Nurses need to know probable difficult parts of the care of obese patients. For instance, they have higher difficulty in urinary catheter and intravenous access placement, have changed social, emotional or psychological necessities, need higher amount of

time for dependent-care requisites. It is very important to perform nursing care and to determine nursing diagnoses and plan nursing care.

CONCLUSION

Obesity is a chronic and relapsing disease due to it's complications and comorbidities. It is a complex disease and a result of genetic, environmental, biological, sociocultural and behavioral factors. Obesity has an increasing prevalence in the world and decrease the quality of life. It is an important health problem in terms of community health due to the related cardiovascular, pulmonary, endocrine, metabolic, psychological and other diseases. Obesity can be treated with surgical intervention and a multidisciplinary team approach. It is very crucial to provide personalized, individualized and holistic nursing care and counseling services to patients who undergo obesity surgery for reducing possible risks in their care and preventing complications and difficulties before, during and after surgery. These services are also important to increase patients' quality of life.

REFERENCES

Alican, F. (2007). *General Surgery Volume II: Surgical Treatment of Morbid Obesity.* Istanbul, Nobel Medical Bookstores.

Angrisani, L., Santonicola, A., Iovino, P., Formisano, G., Buchwald, H. & Scopinaro, N. (2015). Bariatric Surgery Worldwide 2013. *Obes Surg,* Oct 25(10), 1822-32.

Asoglu, O. (Transletor) (2008). Complications of Stomach Surgery (Morbid Obesity). In Yeşim Erbil & Unal Değerli (Translation Editors), *Complications of Surgery* (398-406). Istanbul, Guneş Medical Bookstores.

Barth, M. M. & Jenson, C. E. (2006). Postoperative Nursing Care of Gastric Bypass Patients. *Am J Crit Care*, 15, 378-388.

Bozbora, A. (2002). Obesity Disease. In Goksel Kalaycı (Ed), General Surgery Volume 1. *Istanbul University Istanbul Faculty of Medicine Basic and Clinical Sciences Textbooks* (521-531). Istanbul, Nobel Medical Bookstores.

Frantzides, C. T., Zografakis, J. G., Welle, S. N. (2013). Bariatric Surgery. In Iskender Sayek (Ed), *Basic Surgery Volume 1* (4th Ed, 1333-1341). Istanbul, Guneş Medical Bookstores.

Garza, S. F. (2003). Bariatric Weight Loss Surgery Patient Education, Preparation, and Follow-Up. *Crit Care Nurs Q*, 26 (2), 101–104.

Mulligan, A. T., McNamara, A. M., Boulton, H. W., Trainor, L. S., Raiano, C. & Mullen, A. (2009). *Best Practice Updates for Nursing Care in Weight Loss Surgery. Obesity*, 17, 895–900.

Neil, J. A. (2013). Perioperative Nursing Care of the Patient Undergoing Bariatric Revision Surgery. *AORN Journal*, February 97(2), 210-229.

Organisation for Economic Co-operation and Development (OECD). (2017). *Obesity Update* 2017. Available from: https://www.oecd.org/els/health-systems/Obesity-Update-2017.pdf. (Downloaded July 27, 2018).

Otmalaz, İ. & Ece, İ. (2016). Nursing Care in Obesity Surgery. *Selcuk Medicine Journal,* 32(2), 52-55.

Owers, C. E., Abbas, Y., Ackroyd, R., Barron, N. & Khan, M. (2012). Perioperative Optimization of Patients Undergoing Bariatric Surgery. *Journal of Obesity*, Article ID 781546, 1-6 pages.

Sakrak, O. & Dikmen, K. (Transletor). (2010). Morbid Obesity. In Ali Naki Ulusoy & Koray Topgul (Translation Editors), *Sabıston Textbook of Surgery. Biological Basis of Modern Surgery Practice* (1st Ed, 357-399). Istanbul, Nobel Medical Bookstores.

Tanaka, D. S. & Peniche, A. C. G. (2009). Perioperative Care for Morbid Obese Patient Undergoing Bariatric Surgery: Challenges for Nurses. *Acta Paul Enferm*, 22(5), 618-23.

Turkey Endocrinology and Metabolism Society. (2017). *Obesity Diagnosis and Treatment Guide.* (5th Ed). Ankara, Miki Typography.

Turkish Public Health Institution. (2013). *Struggle with Obesity Handbook*. Publication Number: 904, Ankara. Available from: http://beslenme.gov.tr/content/files/basin_materyal/obezite_mucadele_el_kitabi.pdf. (Downloaded August 07, 2018).

Turkish Statistical Institute. (2015). *Turkey Health Survey*, 2017. Available from: http://www.tuik.gov.tr/basinOdasi/haberler/2017_31_20170607.pdf. (Downloaded August 01, 2018).

World Health Organization. (2018). *Obesity*. Available from: http://www.who.int/topics/obesity/en/ (Downloaded August 07, 2018).

Yurdakul, C. (2015). *The Effect of Patients Nutritional Quality Long Term Clinical and Antropometric Measurements after Bariatric Surgery. Unpublished Master Thesis*. Istanbul Medipol University Health Sciences Institute, Istanbul.

Yuksel, A. (2016). Nutritional Status of a Morbid Obese Patient after Three Years of Bariatric Surgery: Case Report. *Izmir Katip Celebi University Facult of Health Science Journal*, 1(1), 39-45.

In: Advances in Health and Natural Sciences ISBN: 978-1-53614-639-4
Editors: Burcu Yuksel et al.

Chapter 3

EFFECT OF GLOBAL CLIMATE CHANGE ON OCCUPATIONAL HEALTH: HEAT STRESS

Serap Arsal Yildirim
KOÜ Kocaeli Vocational School of Health Services,
Emergency and First Aid Program, Turkey

ABSTRACT

The potential consequences of global climate change have been investigated over the past 20 years, and researchers believed the elderly, newborn, or pregnant people, especially those in the equatorial countries and small islanders are at risk. In addition to these risk groups, another risk group is the workers working outside with brute strength.

The American National Institute for Occupational Safety and Health (NIOSH) reported 423 deaths between 1992 and 2006, mainly due to the rise in temperature and heat waves. Causes of these deaths are often identified as "heatstroke" and other heat-related diseases. It has been determined that 24% of these people were working in forestry, agriculture, fishing, and hunting fields.

The publications between June-July 2018 were scanned through the "PubMed," "CDC" and "WHO" official websites, and the literature on heat stress was examined related the effects of global climate change on occupational health.

This study emphasizes the effect of greenhouse gases emitted uncontrollably and intensely to the atmosphere with the Industrial Revolution on working health as a cause of global climate change.

Keywords: global climate change, occupational health, heat stress

INTRODUCTION

Without greenhouse gases such as carbon dioxide, the earth's temperature would be around -18°C, and it probably wouldn't be possible to talk about life as it is today. With technological improvements, the release of greenhouse gases has also increased, and it has brought a new predicament over the years, global climate change.

Global climate change is making radical changes that cause degradation of ecological balance such as temperature increase, extreme weather phenomena, natural disasters, deterioration of water cycle, change in air quality, increase in sea level, and in human life such as ecological migration, new diseases.

Between the years of 1860-2000, it is recorded that the temperature increased in the world is between 0,5-0,7°C (Brass, 2002). Depending on the climate change caused by human or natural factors, a temperature increase of only 2-4°C is expected in this century alone (Brass, 2002; Alley et al., 2007; Kjellestrom et al., 2011).

Global climate change has direct effects such as an increase in temperature, changes in ozone activity, increase in solar radiation due to the variety of air pollutants, and rise in air moisture content are among the predictions (Grandi et al., 2016). These changes are accompanied by indirect effects such as deterioration of human health. In a report published by the International Panel on Climate Change (IPCC) (2014), among the health problems to be encountered in the future due to the consequences of global climate change, lethal hot strokes, heart attack, drought-related nutritional disorders and infectious diseases, injuries due to extreme air temperatures were found to be equally possible (Smith et al., 2014).

Considering about 1/3 of human life is spent on working, it is one of the expected results that climate change affects work life along with occupational health.

Increased exposure to solar radiation, increase in toxic effects of certain chemical substances due to temperature increase, increase in the number of vectors and pathogens, loss of former value due to the migration of workforce due to migrations, workplace violence or precarious work, work accidents due to extreme temperatures are among the future risks threatening the occupational health and safety. Particularly, the sectors most affected by climate change are the ones that complete the majority or most of their shifts in the outdoor environment or in isolation exposed to extreme temperatures, radiation and high humidity such as agricultural workers, road and construction workers, archaeologists, fishermen, firefighters, disaster emergency workers in fire brigade and ambulance.

METHODS

In this study, the literature review of heat stress related to the effects of global climate change on occupational health was conducted from June to July 2018 using the official publications "PubMed," "CDC," and "WHO" and the written sources in this issue.

Temperature and Physiological Effect

Deaths due to high temperatures in the United States cause almost the same amount of deaths as natural disasters such as earthquakes, lightning strikes, and storms. For this reason, high-temperature fluctuations due to global climate change are considered "silent killers" (Jacklitsch et al., 2016), as they do not leave devastating traces behind like any other natural disasters. While even people in resting circumstances are losing their lives due to high temperatures, it would not be wrong to say that workers who are working with physical strength in unsupervised conditions and not provided with adequate measures are seriously endangered by extreme temperatures.

For the human body to function properly, it is necessary to keep the body temperature persistent. The temperature of the brain, the thorax, the abdomen, which is called the core temperature, is not the same thing as the surface temperature. Core temperature is 37 ± 1°C, whereas the surface temperature is 2-4°C lower (Marchetti et al., 2016). The body uses a thermoregulation mechanism to keep its temperature persistent. In this mechanism, the hypothalamus works as a thermoregulation center, while the frontal lobe reduces the body temperature, and the posterior lobe works to raise the body temperature together with peripheral vessels, striated muscles, and thermoreceptors. The human body can survive in environments at a temperature of 10°C below the core temperature but cannot survive in environments at the temperature of 5°C above the core temperature (Marchetti et al., 2016). For this reason, convection, conduction, radiation, and evaporation develops in the body to stabilize the rising body temperature.

Convection: It is defined as the temperature change between the skin and the ambient air surrounding the skin. If the ambient temperature is larger than 35ºC, body temperature increases. If the ambient air temperature is below 35ºC, body temperature decreases.

Conductivity: The temperature difference between the skin and the surface directly contacting it. In some special CBRN (Chemical, Biological, Radiological and Nuclear Threats) uniforms, furnace work uniforms, liquid circulation special work uniforms, cooling is done with this logic.

Radiation: Heat is displaced by radiation. The body's thermal radiation is usually persistent.

Evaporation: It is defined as heat loss caused by sweat evaporation from the body. The amount of relative humidity in the environment affects the amount of maximum perspiration (Jacklitsch et al., 2016).

Severe physiological changes transpire when the ambient temperature reaches or exceeds the core temperature (Kjellstro et al., 2009; Parsons, 2003). When the body temperature exceeds 38ºC, a decrease in physical and mental abilities is observed, a heat stroke when the temperature exceeds 39°C, damage in large organs, loss of consciousness and death when it exceeds 40,6°C (Marchetti et al., 2016).

The sun exposure in the working environment, the humidity and the temperature of the environment, the worker's uniform, the body surface area, and the worker's job are directly influential in the degree of influence from temperature.

Thermal Comfort

The heat, humidity, airflow conditions are called thermal comfort which is substantial for the most of the workers' relaxation in the working environment. The temperature range of the working environment isn't the only criteria that determines thermal comfort (CDC, 2017), although the temperature range that most people feel comfortable is between 20-26°C. In addition to temperature, humidity is another important criterion. While the very low humidity in the environment can dehydrate the body surface, evaporation does not occur sufficiently under very high humidity conditions. Therefore, the body cannot sweat, and the body temperature cannot be adjusted, it can cause "heat stress" and "hot impact." Ideally, the relative humidity in the working environment should be between 30-80% accordingly the weight of the work done, the season characteristics, the temperature of the environment. The airflow rate is an important factor that causes the exchange of the temperature between the environment and the workers, and the removal of pollutants such as odors and dust in the environment. NIOSH recommends a working airflow rate of 20-150 mm/sec (CDC, 2017; Jacklitsch et al., 2016).

Radiant heat is one of the most important factors that affect working ambient temperature. Radiant heat is electromagnetic waves emitted by radiation from heat and light sources and their heat effect. The solar rays also affect the thermal comfort as solar radiation as well as artificial materials like raw materials produced in the working environment such as glass, molten metal.

Wet Bulb Globe Temperature (WBGT) value is used to determine thermal comfort (Jacklitsch et al., 2016). The WBGT value is the heat load the worker is exposed to in the working environment. In other words, it is the effective temperature that is felt. In other words, it is the effective temperature that is felt. This value is the result of measuring the dry bulb temperature, radiant temperature, and wet bulb temperature in the environment using a formula.

A 1% increase in body temperature leads to a 10% increase in metabolism rate, an increase in the number of pulses, a very slight decrease in systemic blood pressure, and an increase in fluid loss (Yenen, 2017). In addition to these physiological changes, there are risky situations that threaten job security such as careful reduction, fatigue, exhaustion, a decrease in mental functions. Many studies indicate that temperature increase affects the emotional state, prolongs the duration of the reaction, and reduces mental activity in a variety of complex tasks (Lenzuni et al., 2014; Fisk, 2000). While the effective temperature of 29°C leads to a 5% decrease in the work efficiency of the worker, the decrease reaches 30% at 32°C effective temperature.

Heat Stress

The net heat load resulting from the combined contribution of the worker's metabolism, environmental factors, and the working clothing is defined as heat stress (Jacklitsch et al., 2016). This load causes an increase in body temperature by heat storage in the body, which triggers physiological changes. Due to this situation, the respiratory and cardiovascular system may be affected, resulting in acute or chronic insufficiency due to muscle destruction and dehydration (Hansen et al., 2008; Kjellstrom et al., 2010; Blois et al., 2015).

If the heat stress is not prevented, meaning the worker is not taken to rest, if the liquid intake is not provided, and if the person is not cooled, hot cramps, hot syncope and finally heat strokes can be expected. In heat stress, the main factor that determines the answer of the person is the acclimatization, meaning adaptation to work. If the worker is not in acclimatization, he is more affected by the temperature change. If the worker is in acclimatization, the temperature he can tolerate is higher, the duration of endurance is longer without strain. For workers to get the least damage from the heat, NIOSH and ACGIH (Association Advancing Occupational and Environmental Health) regulated workers' working and rest periods according to ambient temperatures (Table 1).

Heat cramps: Basic skeletal muscle cramps is seen because of lack of water and salt replacement lost during perspiration.

Heat syncope: There is an enlargement to cool the body in the peripheral superficial veins, where the water and the salt lost by sweating are not replaced, and the heat exposure continues. This vasodilatation causes blood pressure in the brain to drop and temporary loss of consciousness.

Heat stroke: Working in a hot environment with physical power can increase body heat production from 1 kcal/min to 20 kcal/min. This means that the core temperature increases by 1°C every 5 to 7 minutes (Marchetti, 2016). When exposed to high temperature and moisture for long periods of time, the body uses all means to lost heat. The method of losing most heat is the perspiration. Despite the rising body temperature, sweating on the body may not be enough due to intense moisture in the environment, or perspiration stops because sweat is not replaced. In this case, while the body temperature is above 40°C/104°F the activity of the nervous system over the thermoregulation ceases to exist, and uncontrolled temperature increase and significant tissue damage and death due to this increase can be seen (rhabdomyolysis, thrombocytopenia, coagulopathy and multi-organ failure) (Nag et al., 2009; Parsons 2014). A form of hyperthermia associated with a systemic inflammatory response leading to a syndrome of multi-organ dysfunction in which encephalopathy predominates (Bouchama and Knochel, 2002).

With the global climate change, especially for the workers working in daylight, the risk of heat stress and other heat cramps, syncope and heat stroke risks are increasing, and it is expected that this risk will be higher if the precautions aren't taken (Kjellstrom et al., 2016).

Table 1. NIOSH and ACGIH recommended working/rest periods in hot environment

Recommended work/rest periods	Work load weight*		
	Heavy	Medium	Light
	Environment Temperature (ºC)		
Continuous working	25	26,7	30
75% working 25% rest (for every hour)	25,9	28	30,6
50% work 50% rest (for every hour)	27,9	29,4	31,4
25% working 75% rest (for every hour)	30	31,1	32,2

*: Work Load Weight Assessment: Light jobs: 200 kcal/hour (Example: Writing-like office work), medium jobs: 200-350 kcal /heavy work :> 350 kcal/h (Example: digging works).

Studies in various countries in the world (Table 2) show that extreme heat at work can cause deaths, heat exhaustion, major organ damage, job accidents, and changes the emotional state and attention resulting in a decrease in performance and economic loss.

In a study conducted in Puerto Rico between 2012 and 2013, it is discovered that the highest death rate was recorded between May and October when the death rates within 2 years were compared. The risk of life loss because of illnesses like stroke (brain hemorrhages, clots, paralysis), hypertension has increased between 2 and 3 times in the 2012-2013 Fall in comparison to the 2009-2010 Fall (Mendez Lazaro et al., 2016). In a study conducted by OSHA between 2012 and 2013, 20 workers who had died or suffered from overheating were examined and one of them had died and the others had been exposed to high temperature related illnesses.

More than half of the working hours of workers besides home workers, especially farmers and construction workers, are spent outside. Naturally, the changes in climate characteristics are affecting/will affect the health of this working group, as well as occupational health and safety.

In Paris in 2003, it was seen that most of the women lost their lives because of extreme temperature fluctuations were in high age group, and the males were included in the working group between the ages of 25-64 (Pachauri et al., 2014; Robine et al., 2008). Outside workers will be exposed to high temperatures, radiation, chemical substances and it will threaten the occupational health, mainly because of climate change. In a study conducted by NIOSH, 423 workers lost their lives due to high temperatures between 1992 and 2006, and 102 (24%) of them have been working in forestry, agriculture, fishing, hunting fields (CDC, 2008).

Table 2. Situations that develop because of exposure to high heats in workplaces

Effect	Research Area	Researchers
Heat Stroke	South African Mine Workers, US Agricultural workers, China, India, Qatar and other countries	Tustin et al., 2018; Gibson and Pattison 2014; Nag and Nag, 2009; Grandjean and Grandjean 2007;
Unusual severe heat stroke, heat exhaustion	Some hot workplaces in China, India, England, runners in US and Brasil, footballers in US	Nag and Nag, 2009; Parsons 2014; Zhao et al., 2009; WHO, 2014; Pachauri et al., 2014; Watkins et al., 2018; De Martini et al., 2014; Castro et al., 2018;Kerr et al., 2014.
Kidney damage, Major organ damage	US sugar workers, migrant construction workers in Qatar, agricultural workers in Cameroon,	Garcia Trabanino et al., 2005; Kjellstrom et al., 2010; Tawatsupa et al., 2012; Wesselling et al., 2013, Ekiti et al., 2018
Heat-related accidents	Europe, Thailand	Ramsey et al., 1983; Tawatsupa et al., 2013
Behavioral and psychological changes decrease in mental functions	South African miners, Australian farmers, student children in Thailand, Cameroon	Tawatsupa et al., 2010; Berry et al., 2010; Dapi et al., 2010
Decrease in working capacity, economic loss, migration	India, USA, South Africa	Dell et al., 2009; Burke et al., 2015; Kjellstrom, 2016; Kjellstrom, 2011; Kjellstrom, 2009

(Kjellstrom T., Briggs D., Freyberg C., Lemke B., Otto M., Hyatt O. Heat, Human Performance, and Occupational Health: A Key Issue for the Assessment of Global Climate Change Annu. Rev. Public Health 2016. 37:97-112).

In a study conducted by Eisenberg and Methner (2018), which examined the heat stress conditions of the park workers, didn't detect dehydration and muscle destruction while the decrease in the working efficiency of the people, the increase in the pulse was recorded. Nag and Nag (2009) recorded the most intense heat stress findings (e.g., headache, palpitations, systolic pressure, apathy etc.) in the rock crushing workers working outside among workers working in ceramic, mine (iron), and rock breaking. Tawatsupa et al., (2012) found that, in Thailand, labor-intensive workers were significantly more likely to experience heat stress than workers in the office and that they were experiencing kidney dysfunction due to heat stress, even more in 2009 than in 2005. Gracia Trabanino et al., (2005) emphasized that chronic renal failure and proteinuria in agricultural workers in El Salvador may be due to heat stress. Ekiti et al., (2018) found parallel results in a study conducted by agricultural workers in Cameroon. In a study conducted by Tustin et al., (2018), 78.6% of 25 outside workers in the United States between 2011 and 2016 were exposed to fatal heat stress in acclimatization. Beck et al., (2018) found that WBGT and UV exposures of workers engaged in gardening work were above the legal limits.

It is estimated that in addition to direct health problems of global climate change such as the heat stress, in 2030-2050, particularly in South East Asia and the Ecuadorian belt countries, a decrease in working capacity and a 50% loss in production due to a temperature increase of 2ºC is reported (Kjellstrom et al., 2013). In another study, a modeling study in Southeast Asia has shown that in 2050 more than half of the working hours in the afternoon will be lost due to the greater need for rest breaks. It is estimated that global gross national income will decrease by 23% in the 2100s due to shortening of working hours, reduced work efficiency and other climate-related problems (Kjellstrom et al., 2016).

It is claimed that labor productivity losses will occur in the whole Aegean and Mediterranean coasts in the south of İzmir, and 1% of the losses in the Central Anatolian Region can reach to 7-8% (Altınsoy and Yıldırım, 2013). In a study conducted in our country, a 2001-2050 projection connected to global warming has been made with the labor productivity loss between 1970-2000, and while there is a loss of productivity in the range of 3-18% in İzmir, Aydın, Mersin, and Adana in the past, losses are predicted to rise to 25% in the future.

CONCLUSION

Global climate change is a threat to occupational health due to temperature rise, humidity and extreme warming. In terms of occupational health and safety, the following measures will help to reduce the risk of heat stress and reduce the threat of occupational health and safety; adapting to the NIOSH recommended study/rest periods, taking a 250cc (at a temperature below 15C) fluid every 20 minutes, supporting workers with athletic liquids in case of extreme perspiration, no alcohol in hot weather, wearing air-permeable thin clothes, the presence of a radio on their side, carrying out an air conditioning which can provide thermal comfort in indoor environments, determining working hours not only by the effect of sunlight but also by the temperature of the air for those who work outside for the following years, and frequent control of WBGT values.

REFERENCES

Alley RB, Bernstein T, Bindoff NL, Chen Z, Chidthaisong A, FriedlingsteinP et al., Summery for policymakers. In: Solomon S, Qin D, Manning M, Chen Z, Marquis M, Averyt KB, Tignor M, Miller HL, editors. *Climate Change 2007: The physical science basis. Contribution of Working Group I to the Fourth Assessment Report of the Intergovernmental Panel on Climate Change.* Cambridge, UK and New York, NY, USA: Cambridge University Press, 2007, 10-15.

Altinsoy H, Yildirim HA. *Labor productivity projection of hirham model results in the eastern Mediterranean region*, III. Turkey climate changes congress, TIKDEK 2013, 3-5 Haziran, Istanbul Technical University, Istanbul, 2013.

Bouchama A,, Knochel JP. Heat stroke. *N Engl J Med*. 2002, 346(25), 1978-88.

Blois JD, Kjellstrom T, Agewall S, Ezekowitz JA, Armstrong PW, Atar D. The effects of climate change on cardiac health, *Cardiology,* 2015, 131, 209–217.

Brass GW, *Arctic Ocean Climate Change*, US Arctic Research Commission Special Publication, 02:1, Arlington, Virginia, 2002.

Beck N, Balanay JAG, Jhonson T. Assessment of occupational exposure to heat stress and solar ultraviolet radiation among groundskeepers in an eastern North Carolina university setting, *J Occup Environ Hyg*, 15(2), 105-116.

Berry HL, Bowen K, Kjellstrom T. Climate change and mental health: a causal pathways framework. *Int. J. Public Health, 2010,* 55, 123–32.

Burke M, Hsiang SM, Miguel E. Global non-linear effect of temperature on economic production. *Nature, 2015,* 527, 235–239.

Castro RRT, Filho RC, Lucas Da Nobrega AC. Fulminant liver failure in street runner: Effects of heat stroke, *Rev Assoc Med Bras*, 2018, 64(3), 208-211.

Centers for Disease Control and Prevention (CDC). Heatrelated deaths among crop workers – United States, 1992-2006. *MMWR Morbidity & Mortality Weekly Report* 2008. (2018 June 6). Available from: www.cdc.gov/mmwr/about.html.

Centers for Disease Control and Prevention (CDC). *Heat Related Illness (2018 June 15).* Available from: https://www.cdc.gov/pictureofamerica/pdfs/Picture_of_America_Heat Related_Illness.pdf.

Centers for Disease Control and Prevention (CDC). *Indoor environmental quality. (2018 June 8).* Available from: https://www.cdc.gov/niosh/topics/indoorenv/hvac.html.

Dapi LN, Rocklov J, Nguefack-Tsague G, Tetanye E, Kjellstrom T. Heat impact on schoolchildren in Cameroon, Africa: potential health threat from climate change. *Glob. Health Action*, 2010, 3: doi: 10.3402/gha.v3i0.5610.

De Martini JK, Casa DJ, Belval LN, Crago A, Davis RJ, Jardin JJ, Stearnes RL. Enviromental conditions and occurrence of exertional heat illnesses and exertional heat stroke at the Falmouth road race, *Journal of Athletic Training*, 2014, 49(4), 478-485.

Dell M, Jones BF, Olken BA. Temperature and income: reconciling new cross-sectional and panel estimates. *Am. Econ. Rev. 2009,* 99, 198–204.

Fisk WJ. Health and productivity gains from better indoor environments and their relationship with building energy efficiency. *Annu Rev Energy Environ*, 2000, 25:537-66.

Garc´ıa-Trabanino R, Dominguez J, Jansa J, Oliver A. Proteinuria and chronic kidney disease on the coast of El Salvador. *Nefrologıa,* 2005, 25:31–38.

Gibson O, Pattisson P. 2014. Qatar World Cup: 185 Nepalese died in 2013 - official records. *Guardian*, 2014, Jan. 24.(2018 June 10) Available from: http://www.theguardian.com/world/2014/jan/24/qatar-2022-world-cup-185-nepaleseworkers-died-2013.

Grandi C, Borra M, Militello A, Polichetti A. Impact of climate change on occupational exposure to solar radiation. *Ann Ist Super Sanità,* 2016, 52, (3), 343-356.

Hansen AL, Bi P, Ryan P, Nitschke M, Pisaniello D, Tucker G. The effect of heat waves on hospital admissions for renal disease in a temperate city of Australia. *Int J Epidemiol,* 2008, 37(6), 1359–65.

Jacklitsch B, Williams WJ, Musolin K, Coca A, Kim HJ, Turner N. *NIOSH Criteria for a Recommended Standard Occupational Exposure to Heat and Hot Environments Revised Criteria*, NIOSH Publication: 2016.

Kerr ZY, Marshall SW, Comstock RD, Casa DJ. Exertional heat stroke management strategies in United States high school football, *Am J Sports Med.* 2014, 42(1), 70-77.

Kjellstrom T, Lemke B, Hyatt O. Increased workplace heat exposure due to climate change: a potential threat to occupational health, worker productivity and local economic development in Asia and the Pacific region. *Asian-Pacific Newsletter on Occup Health Safety* 2011, 18(1),6-11.

Kjellstro¨m T, Holmer I, Lemke B. Workplace heat stress, health and productivity-an increasing challenge for low and middle income countries during climate change. *Global Health Action,* 2009, 2, 1-6.

Kjellstrom T, Lemke B, Otto M. Mapping occupational heat exposure and effects in south east Asia ongoing time trends 1980-2011 and future estimates 2050 *Industrial Health* 2013, 51, 56-67.

Kjellstrom T, Butler AJ, Lucas R, Bonita R. Public health impact of global heating due to climate change—potential effects on chronic non-communicable diseases. *Int. J. Public Health*, 2010, 55, 97–103.

Kjellstrom T, Briggs D, Freyberg C, Lemke B, Otto M, Hyatt O. Heat, human performance, and occupational health: a key ıssue for the assessment of global climate change ımpacts, *Annu. Rev. Public Health* 2016, 37, 97-112.

Kjellstrom T. Impact of climate conditions on occupational health and related economic losses: a new feature of global and urban health in the context of climate change. *Asia-Pac. J. Public Health,* 2016, 28(2 suppl.), 28-37.

Lenzuni P, Capone P, Freda D, del Gaudio M. Is driving in a hot vehicle safe? *Int J Hyperthermia,* 2014, 30(4), 250-257.

Marchetti A, Capone P, Freda D. Climate change impact on microclimate of work environment related to occupational health and productivity, *Ann Ist Super Sanità*, 2016, 52, (3), 338-342.

Mendez Lazaro PA, Pérez-Cardona CM, Rodríguez E, Martínez O, Taboas M, Bocanegra A, Méndez-Tejeda R. Climate change, heat, and mortality in the tropical urban area of San Juan, Puerto Rico. *Int J Biometeorol.* 2016, 62(5), 699-707.

Methner M, Eisenberg J. Evaulation of heat stress and heat strain among employees working outdoors in an extremely hot environment, *J Occup Environ Hyg.,* 2018, 15(6), 474-480.

Nag PK, Nag A, Sekhar P, Pandit S. Vulnerability to heat stress: scenario in Western India. *World Health Organ.* (WHO) Rep. APW No. SO 08 AMS 6157206, Natl. Inst. Occup. Health, Ahmedabad, India. 2009.

Pachauri RK, Allen MR, Barros VR, Broome J, Cramer W, Christ R et al., *IPCC Climate Change 2014: Synthesis Report*. Contribution of

Working Groups I, II and III to the Fifth Assessment Report of the Intergovernmental Panel on Climate Change IPCC, Geneva, Switzerland, 2015.

Parsons K. *Human thermal environment - the effects of hot, moderate and cold temperatures on human health, comfort and performance*, 2nd ed. New York: CRC Press, 2003.

Parsons K. *Human thermal environments the effects of hot, moderate and cold temperatures on human health, comfort and performance*, 2nd ed. London: Taylor & Francis. 2014.

Ramsey JD, Burford CL, Beshir MY, Jensen RL. Effects of workplace thermal conditions on safe working behavior. *J. Saf. Res. 1983,* 14, 105–114.

Robine JM, Cheung SL, Le Roy S, Van Oyen H, Griffiths C, et al., 2008. Death toll exceeded 70,000 in Europe during the summer of 2003. *C. R. Biol.,* 2008, 331(2), 171–78.

Smith KR, Woodward A. Human Health: Impacts, adaptation and co-benefits. In: Field CB, Dokken DJ, Mastrande MD, Barros VR, Mach KJ, Bilir TE et al., editors. *IPCC Climate Change 2014,* Impacts, adaptatation and vulnerability. US, Cambridge university press, 2014: 731-741.

Tawatsupa B, Lim LLY, Kjellstrom T, Seubsman S, Sleigh A. Association between occupational heat stress and kidney disease among 37,816 workers in the Thai Cohort Study (TCS). *J. Epidemiol.,* 2012, 22:251–60.

Tawatsupa B, Yiengprugsawan V, Kjellstrom T, Berecki-Gisolf J, Seubsman SA, Sleigh A. Association between heat stress and occupational injury among Thai workers: findings of the Thai Cohort Study. *Ind. Health,* 2013, 51, 34–46.

Tawatsupa B, Lim LLY, Kjellstrom T, Seubsman S, Sleigh A, et al., The association between overall health, psychological stress and occupational heat stress among a large national cohort of 40,913 Thai workers, *Glob. Health Action,* 2010, 3, 5034.

Tustin AW, Lamson GE, Jacklitsch BL, Thomas RJ, Arbury AB, Cannon DL, Gonzales RG, Hodgson MJ, Evaluation of Occupational Exposure

Limits for Heat Stress in Outdoor Workers — United States, 2011–2016, *Weekly*/July 6, 2018, 67, 26, 733–737. (2018 June 15) Available from: https://www.cdc.gov/mmwr/volumes/67/wr/mm6726a1.htm?s_cid=mm6726a1_w#T1_down.

Watkins ER, Hayes M, Watt P, Richardson AJ. Fire service instructors' working practices: A UK survey, *Archives of Environmental & Occupational Health,* 2018, 5, 1-9. DOI: 10.1080/19338244.2018.1461601.

World Health Organisation (WHO). Quantitative risk assessment of the effects of climate change on selected causes of death, 2030s and 2050s. Geneva, *World Health Organization*, 2014. (2018 June 22). Available from: http://apps.who.int/iris/bitstream/handle/10665/134014/9789241507691_eng.pdf, jsessionid=54933CC18E1F12F0AFB5351A6052D628?sequence=1.

Yenen OS. Heat. In: Willke Topçu A, Söyletir G, Doğanay M. Editors. *Infectios disseases and microbiology*. Istanbul, Turkey: Nobel, 2017, 503-528.

Zhao J, Zhu N, Lu S. Productivity model in hot and humid environment based on heat tolerance time analysis. *Build. Environ.,* 2009, 44(11), 2202–7.

In: Advances in Health and Natural Sciences ISBN: 978-1-53614-639-4
Editors: Burcu Yuksel et al.

Chapter 4

PHYSICAL ACTIVITY IN WOMEN

Asuman Saltan*
Therapy and Rehabilitation, Yalova University, Yalova, Turkey

ABSTRACT

Intentional physical activity has become necessary with today's evolving technology. However, individuals need a routine in their lives. Individuals are able to complete a lot of work in their daily lives using technology, but this increases their sitting time. This situation emerges into a sedentary lifestyle. There is a negative relationship between sedentary lifestyle and physical activity. As sitting time increases, physical activity decreases. This situation differs between the genders. One reason for this is that women are still an influential value to society, culture, and the economy. Therefore, women and physical activity need to be examined even in today's conditions.

When women perform physical activity, due to physiological structure, menstruation, pregnancy, hormones, emotional state changes, elastic structures, daily life activities, etc. many factors are involved. For this reason, physical activity in women can have different results at different times. In studies, the tendency toward sedentary behaviors of women are higher than that of men. This section includes evidence-based

* Corresponding Author Email: fzt_asuman@yahoo.com.tr.

information on the implementation and effects of physical activity in women.

Keywords: exercise, physical, women, sedentary lifestyle

INTRODUCTION

Human survival depended on being physically active and talented in ancient times. People had to live a nomadic lifestyle to meet basic needs such as food and shelter. Thus, they frequently ran, jumped, fought, and walked long distances, thereby performing excessive physical activities. After many centuries, a decrease in physical activity rates occurred. Although the number of cultural and social classes has decreased compared to the past, individuals have begun to be inactive after the developments in technology and industry (Brown, 2016).

Today, the definition of physical activity has changed. Physical activity is now defined as entire body movements during which muscles work with a level of energy that is higher than the metabolism relaxation level. This definition includes four headings: 1. Physical activities during leisure time, 2. Physical activities in the workplace, 3. Physical activities during housework, and 4. Physical activities in society (Brown, 2016; Timperio, 2015).

This classification may vary by country. For example, Brazilian public health studies report that physical activities at home or in the workplace should not be calculated. In Latin America, upper limits are exceeded when physical activities at home and in the workplace are calculated. Therefore, physical activities are categorized into two groups: leisure time activities and social physical activities. These differences between countries are important and should be taken into consideration during academic studies. In particular, information should be obtained from the relevant public health departments regarding the definitions of physical activity (Brown, 2016; Timperio, 2015).

Physical inactivity is a life-threatening factor around the world according to the World Health Organization (WHO) that particularly affects women. Approximately 3.2 million people die because of physical inactivity every year. Regular physical activity and exercise protects health and physical fitness and prevents many health problems. Previous studies report strong effects of regular and evidence-based physical activities and exercises on health (Arslan, 2016; Brown, 2016; Timperio, 2015).

Women's health state is dynamic because their hormonal levels are dynamic. Changes in the osseous-muscular organs in non-disease periods such as menstruation, pregnancy, or menopause affect women's health state and quality of life. Thus, the target level of physical activity varies by the phase of the woman's life. In addition, the effect of sociocultural and environmental factors makes women's daily lives complicated (Arslan, 2016; Brown, 2016; Timperio, 2015; Cooke, 2017). This paper aims to reveal the details on the relationship between physical activity and women's health.

Exercise and Bone Health in Women

Bone health plays an important role in health and the quality of life. Bone tissue is home to minerals in addition to its functions including preventing injuries, protecting bodies, and moving. Thus, it contributes to the normal functions of other organs (from regulating the functionality of endocrine organs to hematopoiesis). An unhealthy bone may lead to breaks and functional losses, which result in disability. This may lead to undesired results including functional dependency and premature death. Osteoporosis is a bone health issue widely observed in women. Many factors affecting bone strength such as changes in bone tissue, decreases in bone density, and changes in bone architecture emerge during osteoporosis. As a result, bones can break easily. For example, one in every two Caucasian women is diagnosed with osteoporosis and osteoporotic fractures are widely observed (Xu, 2016; Zhou, 2014). Although one in every five men also experience osteoporotic fractures, women are at greater risk during their lifetime. Postmenopausal women have increased expenses for synthetic drugs used in the treatment of their higher osteoporosis prevalence. Expenses for elderly women in England are expected to reach 25.3 million dollars by the year 2025 (Herrmann, 2012; Xu, 2016). Thus, an increase in non-pharmacological interventions is desirable. Exercise prevents breaks and falls in postmenopausal and elderly women, reduces or stabilizes bone loss in middle-aged women, and increases bone density in girls. The WHO, American Collage of Sport Medicine (ACSM), National Institutes of Health (NIH), and National Osteoporosis Foundation (NOF) provide suggestions for the treatment and prevention of osteoporosis. However, these are general suggestions that are not valid for specific age groups. High quality systematic reviews and meta-analyses suggest that exercises done for bone health of women from diverse age groups have positive effects (Xu, 2016).

Effect of Exercise on Girls

School practices consisting of short-term plyometric exercises are adequate for the bone health of girls. Exercise programs for 60 minutes or more three days a week or 10- to 20-minute programs performed daily for 12 months or longer are effective for bone development. These programs will be particularly useful for girls with precocious puberty. Awareness should be raised in schools to enable young girls to have healthy bone development, and emphasis should be placed on the effect of school-based exercise programs on bones (Ubago-Guisado, 2015; Xu, 2016).

Effect of Exercise on Premenopausal Women

Studies show that regular physical activities have a limited effect on the bones of premenopausal women. However, regular exercises have an effect on the femur neck, lumbar vertebra, and trochanter. In this period, the aim is not to increase bone strength as in girls, but to protect strength and prevent it from decreasing. Resistance exercises tolerated or performed by individuals with moderate to high effects are preferred for bone health during this period. Exercises performed 30- to 60-minutes at least three days a week for 10 months or longer are effective for maintaining bone health in premenopausal women (Xu, 2016).

Home or office exercises are sought for women who do not have enough time for traditional exercises due to familial matters or workload (Xu, 2016). Short-term, high-volume vertical jumping or hopping performed in 1 to 5 sets of 10 to 20 repetitions at 8- to 15-second intervals at least three times a week for six months are recommended (Xu, 2016; Farr, 2013).

Effect of Exercise on Elderly Postmenopausal Women

Because the prevalence of osteoporosis is higher in the postmenopausal period, the importance of exercises for bone health remains throughout aging. Combined exercise methods are frequently recommended in this period. Exercises lasting for 30- to 60-minutes for a minimum of three days a week for more than 10 months are recommended. Recommended exercises include running, steps, plyometric exercises, jogging forward or backward, etc. Performing these exercises together protects the structure of bone tissue (Xu, 2016; Chastin, 2014).

Exercise in the Pregnancy Period

Many women hesitate to perform exercises during the pregnancy period. Physiotherapists recommend women who were physically active before pregnancy reduce their activity levels. However, exercises have been optimized for pregnant women, as resources from the last 30 years show. Today, physically active pregnant women perform their activities in the same way as they did before pregnancy. Even those who were not active before pregnancy tend to increase their physical activities during pregnancy (Bredin, 2013).

In the past, the primary goal was to protect the fetus and the mother. At that time, people thought that the fetus could be harmed by several factors such as energy substrates or increased body temperatures in addition to the increased oxygen required by the working muscles of the mother. Furthermore, performing too much exercise during the pregnancy period was considered risky because it may increase the likelihood of injuring the lower back or the musculoskeletal system, such as chronic fatigue or hypoglycemia. Therefore, many physicians suggested reducing physical activities during pregnancy (Bredin, 2013).

Systematic reviews between 1982 and 2009 show that exercises performed by pregnant women with no contraindications are only slightly risky (Bredin, 2013). Uterine contractions that can last for 72 hours are the leading side effect. However, previous studies show that fetal bradycardia is the most severe side effect. On the other hand, leg cramps, fatigue, and weakness are defined as temporary problems rather than side effects (Bredin, 2013).

If women want to exercise during pregnancy, a health care professional should accompany them. Health care professionals should examine the general health of the mother before preparing the exercise program. They should first eliminate the risks that have side effects. Then, they should prepare a program that will last at least 30 minutes in a day, three days a week during the second trimester with a level that is 50-70% of the maximum heart rate. Water activities such as steady cycling, walking, and swimming are among recommended exercises (Bredin, 2013; PAR-Q+ Collaboration (PAR-Q+), 2011; PAR-Q+ Collaboration (ePARmed-X+), 2011). Bredin et al. (2013) suggested professionals design a flow chart similar to Figure 1.

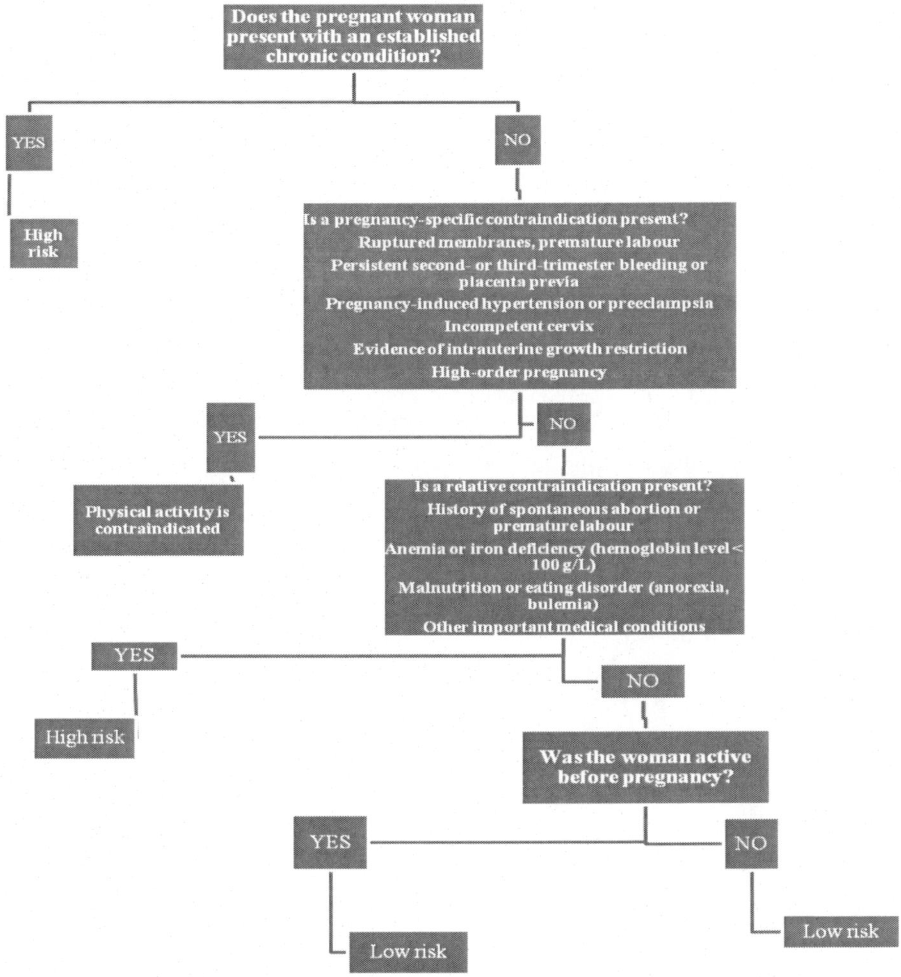

Figure 1. Professionals design for assessing the risk of adverse events during physical activity in pregnant women at clinic (Bredin, 2013).

Medium-level exercises have no side effects or they have only slight side effects for healthy mothers or their children if this is the mothers' first pregnancy and there are no complications. Being inactive before pregnancy is not a reason for not doing exercise during pregnancy. If the examination conducted by a health care professional shows a pregnant woman is at low risk, there is also a low likelihood for the emergence of side effects from the exercises. A screening process should certainly be conducted before calculating the rate of physical activity or starting any activity program (Bredin, 2013; PAR-Q+ Collaboration (PAR-Q+), 2011; PAR-Q+ Collaboration (ePARmed-X+), 2011).

Although regular physical activity is recommended during pregnancy, pregnant women consider that loss of balance and joint or connective tissue traumas may occur. They sometimes avoid performing an activity for fear of causing trauma to the fetus. The most significant point here is that pregnant women should exercise with a health care professional (such as a physiotherapist) (Bredin, 2013).

Thus, the question "Does the emergence of side effects from regular physical activity among inactive or active women during pregnancy differ by fetus or mothers' themselves?" which was specified in the previous literature is still valid, and there is a need for further research. This study implies that pregnant women with no contraindications should be directed to medium-level activities. The relevant literature does not provide precise information about the level of physical activity acceptable during the pregnancy period. For example, whether sportswomen can continue performing their activities at the same intensity level during pregnancy is still unknown (Bredin, 2013; Cooke, 2017).

Studies conducted between 1950 and 2015 indicate that exercise has no side effects. A study conducted in the 1960s reported that being inactive during the pregnancy might lead to negative outcomes such as premature labor, miscarriage, or placental abruption (Bredin, 2013). The application method of exercises is not clear during the pregnancy of sportswomen with high levels of physical activity (Bredin, 2013; Cooke, 2017). Future studies on exercise during pregnancy should question if women's exercise level depends on factors such as education and socioeconomic levels. However, exercise during pregnancy has positive physical and psychological effects. Women's physical activity level before the pregnancy and the sports played before pregnancy should be questioned. The upper limit of the exercise level during pregnancy is still unknown for women who play sports or have high levels of physical activity.

WOMEN AND SPORTS

Previous studies report an increase in the rate of women who play sports. The Lancet Series published in 2016 found a relationship between public health and physical activity starting from the 2012 London Olympics to the 2016 Rio Olympics. In the 2020 agenda of the International Olympic Committee, a participation rate of 50% for women and girls in sports activities is targeted (Cooke, 2017). Participation of women in the Olympics should be encouraged because men fill the open positions because they participate in more sports than women do (Sallis, 2016). Interestingly, the universal estimations of the WHO report the rate of physical activity except sports among women is 35% higher than among men (WHO Global Health Observatory, 2018). Here, the definition and types of physical activity should be closely examined. In addition to leisure time activities, workplace activities or physical activities performed during travel should be considered while preparing the reports. Thus, women may be considered more physically active than men, particularly in countries with low-income levels because women have higher workloads than men do. However, the mean score for women's leisure time activity was 13.7 minutes/day whereas this rate was 25.3 minutes/day among men in low- or middle-income countries (Cooke, 2017).

Participation in sports increases leisure time activities among women. It not only improves physical health, but also mental health and social participation. The UN Inter-Agency Task Force on Sport for Development and Peace reported that as the number of people participating in sports activities increases, the number of physically active individuals also increases. Additionally, the rates of education, maintenance of peace, reduction of trauma, and economic development also increase (Brown, 2016; Cooke, 2017).

The rate of participating in physical activities among different countries varies by gender, age, disability, education level, and socioeconomic status. However, while sports are the fastest way to enable individuals to lead an active life, there is no specific method to increase participation in sports activities. This is apparently due to the presence of different methods performed during the activities (Cooke, 2017). Gender differences in sports participation occur in low- or middle-income countries due to several factors such as cultural/religious values and the role of women in the society. For example, countries such as Qatar, Saudi Arabia, and Brunei had teams of women who participated in the 2012 Summer Olympics for the first time dressing in accordance with their beliefs (Brown, 2016; Cooke, 2017).

WOMEN AND E-HEALTH

Previous studies show physical activity is a highly modifiable health behavior. The WHO suggests 150-minutes of medium-level aerobic activities per week. Although regular physical activity is useful for many conditions from hypertension to depression, the rate of physical activity among women ranges from 3% to 14% (Reed, 2015).

Modern life has placed many additional burdens on women who work outside the home. Women try to exist in every facet of life including the workplace, home, and social areas. High levels of physical activity do not occur during childcare, cooking, gardening, and house cleaning. Thus, women's participation rate in physical activities is 22 to 33% lower than men's is. In low-income families, women even tend to find a second job. Many women report that they too busy to spare time for sports. In this case, e-health services promote physical activities and facilitate their follow-up (Qualcomm Wireless Reach, 2014; Reed, 2015; World Health Organization, 2014).

In a general sense, e-health service is the use of emergency information and communication technology. Previous studies found that e-health services are effective for fitness, nutritional practices, weight loss and conservation, management of chronic diseases and medications, psychological factors, health risks, and health care expenses. E-health services have allowed an increase in the rate of physical activities among women and raised awareness of health interventions such as quality of life, blood pressure, and glucose levels. E-health allows individuals to see their electronic health records when required (Qualcomm Wireless Reach, 2014; Reed, 2015; World Health Organization, 2014).

E-health services in developing countries include age, education level, number of children, work status (part time, full time, or casual work), country, income level, physical activity level, type of e-health service (short message or online web form, virtual care programs, phone or smart phone applications, or gathering data from wearable physical activity measurement instruments), and intervention mode (wearable, telemedicine, mobile devices, health informatics, or multi-single component). However, further studies are required to determine which e-health service is most effective and how it increases physical activities because e-health services have not been included in the relevant literature, yet.

REFERENCES

Arslan, E., Aras, D., Can, S. (2016). Sporcu ve sedanter kadınlarda günlük enerji harcaması ve fiziksel aktivite düzeylerinin karşılaştırılması [Comparison of daily energy expenditure and physical activity levels in sportsmen and sedentary women]. *Ankara Üniv Spor Bil Fak,* 14 (1), 53-61.

Brown, WJ., Mielke, GI., Kolbe-Alexander, TL. (2016). Gender equality in sport for improved public health. *The Lancet,* 388(10051):1257-1258. DOI: https://doi.org/10.1016/S0140-6736(16)30881-9.

Bredin, SSD., Foulds, HJA., Burr, J. F., Charlesworth, S. A. (2013) Risk assessment for physical activity and exercise clearance. In pregnant women without contraindications. *Canadian Family Physician*, 59(5):515-517.

Chastin, S. F., Mandrichenko, O., Helbostadt J. L., Skelton, D. A. (2014) Associations between objectively-measured sedentary behaviour and physical activity with bone mineral density in adults and older adults, the NHANES study. *Bone*, 64:254-62. doi: 10.1016/j.bone.2014.04.009. Epub 2014 Apr 13.

Cooke, R,. Jones, A. Recruiting adult participants to physical activity intervention studies using sport: a systematic review. *BMJ Open Sport Exerc Med*, 2017;3:e000231. doi:10.1136/bmjsem-2017-000231.

Farr, J. N., Laddu, D. R., Blew, R. M., Lee, V. R., Going, S. B. (2013) Effects of physical activity and muscle quality on bone development in girls. *Med Sci Sports Exerc,* Dec;45(12):2332-40. doi: 10.1249/MSS.0b013e31829c32fe.

Herrmann, D., Hebestreit. A., Ahrens, W. (2012) Impact of physical activity and exercise on bone health in the life course: a review. *Bundesgesundheitsblatt Gesundheitsforschung Gesundheitsschutz*, 55(1):35-54. doi: 10.1007/s00103-011-1393-z.

PAR-Q+ Collaboration. (2011). *The Physical Activity Readiness Questionnaire for Everyone (PAR-Q+). Vancouver, BC: PAR-Q+ Collaboration*; (Accessed 2018 Apr 9). Available from: www.eparmedx.com.

Qualcomm Wireless Reach, Vital Wave, GMSA. *Transforming Women's Livelihoods through Mobile Broadband*. 2014.

Reed, J. L., Prince, S. A., Cole, C. A., Nerenberg, S., Tulloch, H. E., Fodor, J. G., Agnieszka Szczotka, A., McDonnell, LA., Mullen, KA., Pipe, AW., Reid, RD. (2015). E-health physical activity interventions and moderate-to-vigorous intensity physical activity levels among working-age women: a systematic review protocol. *BMC Systematic reviews journal*, 4(3):3-7.

Sallis, J. F., Bull, F., Guthold, R. (2016). Lancet Physical Activity Series 2 Executive Committee. Progress in physical activity over the Olympic quadrennium. *The Lancet* (published online July 27.) http://dx.doi.org/10.1016/S0140-6736(16)30581-5.

Timperio, A., Veitch, J., Carver A. (2015). Safety in numbers: Does perceived safety mediate associations between the neighborhood social environment and physical activity among women living in disadvantaged neighborhoods. *Prev Med*, 74:49-54.

Ubago-Guisado, E., Gómez-Cabello, A., Sánchez-Sánchez, J., García-Unanue, J., Gallardo, L. (2015). Influence of different sports on bone mass in growing girls. *J Sports Sci,* 33(16):1710-8. doi: 10.1080/02640414.2015.1004639.

WHO Global Health Observatory. Prevalence of insufficient physical activity. (accessed April 10, 2018) http://www.who.int/gho/ncd/risk_factors/physical_activity/en/.

World Health Organization. eHealth and innovation in women's and children's health: a baseline reivew based no the findings of the 2013 survey of CoIA countries by the WHO Global Observatory for eHealth. *World Health Organization*; 2014. p. 1–164.

Xu, J., Lombardi, G., Jiao, W., Banfi, G. (2016), Effects of Exercise on Bone Status in Female Subjects, from Young Girls to Postmenopausal Women: An Overview of Systematic Reviews and Meta-Analyses. *Sports Med.* 46(8):1165-82. doi: 10.1007/s40279-016-0494-0.

Zhou, B., Wang, J., Stein, E. M., Zhang, Z., Nishiyama, K. K., Zhang, C. A., Nickolas, T. L., Shane, E, Guo, E. (2014). Bone density, microarchitecture and stiffness in Caucasian and Caribbean Hispanic postmenopausal American women. *Bone Res,* 2: 14016. https://doi.org/10.1038/boneres.2014.16.

In: Advances in Health and Natural Sciences ISBN: 978-1-53614-639-4
Editors: Burcu Yuksel et al.

Chapter 5

THE ENHANCED RECOVERY AFTER SURGERY PROTOCOL AND THE ROLE OF SURGICAL NURSES

Selda Mert Boğa*
Vocational School of Health Services,
Kocaeli University, Kocaeli, Turkey

ABSTRACT

The protocols were established to reduce the length of hospital stay of patients because the healing of patients is negatively affected due to long hospitalization as a result of traditional practices. Enhanced Recovery After Surgery Protocol (ERAS) is a evidence-based care model which was designed to accelerate post-operative healing and provide recommendations for patient care applications to be performed before, during and after the surgery. In the protocol that requires a multidisciplinary approach, surgical nurses play an active role at each stage of the process. Therefore, surgical nurses should be aware of their dependent and independent roles in the ERAS protocol, reflect them in the care of the patient, follow the current publications about the protocol and be open to develop. In this chapter, the ERAS protocol components, recommendations and the role of the nurses in the surgical process are emphasized.

Keywords: enhanced recovery after surgery protocol, fast track surgery, surgical nurse

INTRODUCTION

It is known that healing after surgery can be accelerated and surgery-related deaths can be reduced by the current approach for treatment and care and evidence-based applications in surgery. One of the evidence-based applications in this direction is the Enhanced Recovery After Surgery (ERAS) or Fast Track Surgery (FTS) protocol developed by the ERAS Society (Abdikarim et al. 2015). The ERAS program is evidenced-based protocols which were designed to standardize and optimize perioperative medical care in order to reduce surgical trauma, perioperative physiological stress and organ dysfunction related to elective procedures. In addition, improved outcomes, decreased hospital length of stay and faster patient recovery to normal life are expected to be obtained. Other advantages of this philosophy are the reduction of clinical complications and the health costs together with and increase of patient satisfaction (Jimenez et al. 2014, p.487).

The ERAS protocol is a combination of evidence based findings which works synergistically to accelerate the post-surgical recovery period, contains recommendations for patient care at different stages of the operation and requires multidisciplinary (surgeons, anesthetists, nurses, physiotherapists and dietitians) approaches (Kehlet and Wilmore 2005, Varadhan et al. 2010, Demirhan and Pınar 2014, Neville et al. 2014, Karen et al. 2015, Çilingir and Candaş 2017, Watson 2017).

The main goals of the protocol is to reduce postoperative morbidity and mortality, to reduce the duration and cost of hospital stay and to affect patients' perceptions about surgery positively by reducing the reaction of the body to surgical stress via optimizing perioperative nutritional status and promoting analgesia without opioids and postoperative early nutrition (Ljungqvist et al. 2017, Karen et al. 2015, Thorell et al. 2016, Çilingir and Candaş 2017, Yang et al. 2017, Aksoy et al. 2018, Li et al. 2018, Persico et al. 2018). In addition, it is aimed to minimize the pain, complications and functional disorders that may occur after discharging, with the ERAS protocol (Jimenez et al. 2014, Hübner et al. 2015, Çilingir and Candaş 2017). The success of the ERAS protocol depends on the preferably complete fulfillment of the protocol; nurses are an important part of this program. The components and recommendations of the ERAS protocol and the role of surgical nurses before, during and after the surgery are described below.

PREOPERATIVE PERIOD

Preadmission Information, Education and Counselling

Perioperative education for patients and their family members is essential for successful navigation through the surgical experience. There are limited studies that address the benefits of education before surgery; however, evidences show that preoperative teaching conducted before hospital admission is mostly successful. Ronco et al. (2012) found that patient information provided before surgery and reviewed after admission is more effective than information solely received before surgery. After the plan for surgery is confirmed, the patient should begin to receive information from the medical team. The education that the patient receives in the office will set the tone for the surgical experience. The goal of an effective patient education plan is to inform the patient and his or her family members about what to expect during each phase of the experience. Patients often worry about how long they will be hospitalized, types of dressings, driving restrictions, and when they can return to work and household activities, among other things. The goal of the education is to address these stresses and to help patients adequately prepare for and anticipate needs during recovery. A comprehensive education plan helps patients and caregivers engage in the process to prepare for the surgical experience. Including caregivers in the education process helps minimize stress and provides patients and their family members with more realistic expectations to better prepare for returning to home (Karen et al. 2015).

Optimisation of the Patient's Health Before Surgery

It is emphasized that smoking and alcohol use should be quit four weeks before surgery. Alcohol users are at high risk for wound healing, cardiopulmonary function and bleeding; mortality rates are 2-3 times higher in alcohol users (Gustafsson et al. 2013). Pulmonary function and wound site complications are more frequent in smokers (Tunç Tuna and Kurşun 2018).

Preoperative Colon Cleansing

Mechanical colon cleansing is traditionally thought to alleviate the severity of sepsis in the event of anastomotic leakage. However, recent studies have shown that mechanical intestinal cleansing which was started to be applied in 1887, caused stress, dehydration, electrolyte imbalance, long-lasting hunger and delayed normalization of bowel functions; new methods are used instead of this application (Aksoy et al. 2018).

Preoperative Fasting and Carbohydrate Therapy

Stopping oral solid and liquid food intake as from the midnight before elective surgery applications, has been started to apply to reduce the possibility of pulmonary aspiration; it is applied daily as a routine today. However, it has been reported that there is a decrease in the well-being of the patient and some metabolic adversities, especially post-operative insulin resistance after this application (Ersoy and Gündoğdu 2007; Demirhan and Pınar 2014). Today, the accepted fasting protocol is to consume clear liquid food for up to two hours before surgery and to consume solid foods for up to six hours before surgery (Ersoy and Gündoğdu 2007; Yang et al. 2017). In addition, it has been stated that giving 800 ml of carbohydrate fluid until midnight before surgery and giving 400 ml of carbohydrate fluid until 2-3 hours before surgery, accelerate healing after surgery, reduces insulin resistance and significantly shortens the length of hospital stay (Ersoy and Gündoğdu 2007, Demirhan and Gündoğdu 2007, Demirhan and Pınar 2014, Aksoy et al. 2018). It has been suggested that diabetic drugs can be given in combination with carbohydrate therapy in diabetic patients (Gustafsson et al. 2013). In this period, nurses should ensure fasting of patients in the pre-operative period and apply medications.

Tromboembolism Proflaxis

In patients having symptoms, it has been foreseen to use it in appropriate doses (Ersoy and Gündoğdu 2007). In addition, thromboembolism preventive compression socks have been recommended (Demirhan and Pınar 2014).

Antimicrobial Prophylaxis

Antibiotic prophylaxis is used to reduce wound infections after surgery. The antibiotics affecting both aerobic and anaerobic organisms should be administered only once (30 min) before the surgical incision (Demirhan and Pınar 2014, Yang et al. 2017). It has been recommended to repeat a second dose in prolonged surgery (longer than 3 hours) (Ersoy and Gündoğdu 2007, Demirhan and Pınar 2014).

Premedication

The use of antiarrhythmic drugs, beta blockers and statins is also appropriate if there is an indication for premedication (Aksoy et al. 2018). It has been not recommended to administer long or short-acting sedative drugs before surgery because it will delay the healing after surgery (Gustafsson et al. 2013, Solak-Kabataş and Özbayır, 2016).

Intraoperative Period

Anesthesia Protocol

In the ERAS, it is aimed to reduce the surgical stress response. The anesthetic technique should be applied in a way to allow the patient to recover quickly and minimize the opioid effect during the recovery period (Demirhan and Pınar 2014, Solak-Kabataş and Özbayır 2016). The application of the rapid onset and short-acting hypnotic anesthetics, opiates and muscle relaxants shortens the healing period in the postoperative period (Ersoy and Gündoğdu 2007, Demirhan and Pınar 2014). If possible, thoracic epidural analgesia should be used in laparoscopic colon surgery. The application of anesthesia and analgesia via epidural catheters in colorectal surgery reduces insulin resistance, increases motility via preventing sympathetic inhibition of the gastrointestinal tract and makes the use of narcotic analgesics unnecessary (Ersoy and Gündoğdu 2007, Solak-Kabataş and Özbayır 2016).

Selection of Surgical Incisions

The use of transverse or oblique incisions in abdominal surgery can reduce pain and pulmonary risks. A vertical incision can be preferred if adequate vision is achieved. According to the ERAS protocol, it is not a definite binding rule but it is necessary to use the shortest possible time (Demirhan and Pınar 2014).

Nasogastric Intubation

The ERAS protocol has argued the idea that drains, tubes and catheters should be used limitedly and removed immediately after surgery even if they were placed during the operation (Ersoy and Gündoğdu 2007, Demirhan and Pınar 2014, Aksoy et al. 2018). It has been stated that the use of nasogastric intubation does not alter the rate of bowel function recovery, reduce pulmonary complications and hospital stay or other postoperative complications but it delays oral feeding and increases the incidence of pneumonia, atelectasis and fever (Aksoy et al. 2018).

Use of Drains

Postoperative draining has not been recommended because it may prevent the mobility of the patient (Demirhan and Pınar 2014, Solak-Kabataş and Özbayır 2016). It has also been stated that the use of drains has no effect on anastomotic leaks (Ersoy and Gündoğdu 2007).

Urine Catheter

Bladder catheter should be taken early because of disadvantages such as urinary infection and mobility restriction. However, the probability of urinary retention increases as a result of epidural blockage, it is necessary to keep the catheter as long as the blockage continues (Ersoy and Gündoğdu 2007, Demirhan and Pınar 2014).

Perioperative Fluid Management

In modern fluid therapy, fluid infusion which is enough to keep the patient as hydratred, is accepted as adequate. The hyperhydration in intraoperative and early postoperative periods (especially NaCl containing fluids) has been reported to increase the complication rate and the length of stay in the hospital (Ersoy and Gündoğdu 2007). The fluids (colloids and crystalloids) should be administered at the optimal level thereby monitor the cardiac function of the patients during the operation. It is aimed to decrease the length of hospital stay of patients and to improve gastrointestinal function of patients in a short time (Solak-Kabataş and Özbayır 2016).

Prevention of Intraoperative Hypothermia

Hypothermia that occurs during surgery stimulates sympathetic discharge and metabolic endocrine response, increases bleeding via disrupting clotting balance, causes postoperative dizziness and also increases pain at the incisional site via increasing oxygen consumption. Nowadays, it is recommended to start to warm up before the surgery to reduce the central and peripheral temperature differences of the patients and to ensure normothermia (Aksoy et al. 2018). In the literature, it has been reported that this practice reduces blood loss, wound infection, unwanted cardiac events, patient discomfort, the starting time of intestinal motility and the duration of hospitalization (Alfonsi et al. 2014, Aksoy et al. 2018).

It has been recommended that heaters should be used to prevent hypothermia during surgery; the IV fluids should be above 36°C to maintain the body temperature (Gustafsson et al. 2013, Solak-Kabataş and Özbayır 2016). The heating should be continued until the patient's body temperature increases up to 37°C. Active heating should be stopped when body temperature is above 37°C (TARS 2013). It has also been shown that a 1,90°C decrease in body temperature increases the risk of wound infection and the duration of hospitalization by 20% (Mehta and Barcley 2014). Maintaining body temperature at normal levels is one of the critical components of the care before, during and after surgery. Nurses should prevent the frequency of complications in patients by accurate follow-up of patients' body temperature (Oshvandi et al. 2014).

POSTOPERATIVE PERIOD

Elimination of Nausea and Vomiting After Surgery

Postoperative nausea and vomiting are important determinants of well-being and should be prevented because they limit oral feeding of patients in early period (Ersoy and Gündoğdu 2007). The prevention or reduction of nausea and vomiting can reduce fluid electrolyte imbalance, wound dehiscence, haemorrhage, the risk of aspiration pneumonia, the duration of hospitalization and re-hospitalization rates (Miles et al. 2015). The first strategy to reduce nausea and vomiting after surgery is antiemetic prophylaxis. The second strategy is to avoid the use of general anesthesia, propofol and nitric oxide and to reduce opioid use for reducing nausea and vomiting after surgery (Aksoy et al. 2018).

Stimulation of Gastrointestinal Motility

Stimulating gastrointestinal motility and more importantly, avoiding from the use of agents that adversely affect motility is essential for the maintenance of early enteral feeding (Yang et al. 2017). For this purpose, it has been recommended to use epidural analgesia, to avoid the use of opioids, excessive hydration and to use oral magnesium oxide 2x1 gr/day (Ersoy and Gündoğdu 2007, Demirhan and Pınar 2014).

Postoperative Fluid Intake

Oral fluid intake should be started as early as possible after surgery and patients should be encouraged to consume fluids (Yang et al. 2017). High-energy fluids contribute to the normal healing process because of the protein/carbohydrate in their contents. In addition to vitamins, minerals and trace elements, it is recommended that carbohydrates be prescribed as 150 kj/100 ml and proteins as 3-6 g/100 ml. Balanced crystalloid solutions such as 0.9% sodium chloride to prevent hyperchloremic acidosis. The dose and application duration of starch solutions should be limited due to side effects such as renal dysfunction and bleeding. The urine output is the first symptom of hemorrhoids after surgery and therefore oliguria should be evaluated well by nurses. The medical examination of the lungs, neck and veins, heart, abdomen, and extremities provides information about the volume (Aksoy et al. 2018).

Postoperative Analgesia

Optimal control of the pain along with reducing opioid use in surgery as much as possible, is one of the most important components of the ERAS protocol (Aksoy et al. It has been recommended that the postoperative pain control should be performed without the use of opioid analgesics as much as possible because they negatively affect gastrointestinal system, respiratory functions and central nervous system (Solak-Kabataş and Özbayır 2016, Yang et al. 2017).

The long-continued studies on the multimodal analgesic approach proposed by the ERAS protocol have shown that the use of non-steroidal anti-inflammatory (NSAID) drugs (acetaminophen, gabapentin and dexamethasone) reduces pain and minimizes opioid use (Khan et al. 2009, Thorell et al. al. 2017, Aksoy et al. 2018).

Early Removal of Nasogastric Tube, Foley Catheter, and Drainage Tube(s)

All indwelling catheters and tubes will potentially increase the incidence of infection and sepsis, although most ERAS protocols and expert consensus recommend minimizing the use of indwelling catheters and tubes. If medically needed, those inserted catheters and tubes should be removed as soon as medical conditions allow. For patients needing indwelling centralvenous catheterization, the catheter should be removed and replaced in no >1 week or whenever the patient's body temperature escalates (Yang et al. 2017).

Postoperative Nutrition

One of the most important criteria for postoperative well-being and early discharge is early oral feeding (Ersoy and Gündoğdu 2007). Postoperative intravenous fluids should be prematurely cut; oral feeding should be started as soon as possible. Early postoperative oral feeding has been shown to reduce postoperative complications, mortality and the length of hospital stay (Solak-Kabataş and Özbayır 2016). Patients should be encouraged to drink liquid at 2nd hour after surgery and to eat solid food at 4th hour after surgery. Oral nutrition solutions should be provided as a supplement until the maintenance of adequate oral nutrition. Postoperative nutritional support should be continued for at least 8 weeks in patients taking preoperative nutritional support due to their impaired nutritional status (Ersoy and Gündoğdu 2007, Demirhan and Pınar 2014).

Early Mobilisation

Post-operative bed rest is not desirable because it causes to lose muscles, to decrease muscle strength, respiratory function and tissue oxygenation, venous stasis and thromboembolism (Demirhan and Pınar

2014, Yang et al. 2017, Aksoy et al. 2018). According to the ERAS protocol, patients should be kept out of bed for 2 hours on the operation day and for 6 hours a day until the discharge on the following days (Ersoy and Gündoğdu 2007, Khan et al. 2009, Tunç-Tuna and Kurşun 2018). One of the responsibilities of nurses is to help early ambulation (Çam 2015, Tunç-Tuna and Kurşun 2018).

Duration of Hospital Stay and Discharging Criteria

Since the duration of hospital stay is shortened as the primary outcome of the ERAS, planning of discharge should be emphasized both during preoperative patient informing program and hospitalization. A careful and detailed discharge planning should be done to reduce the number of re-applications to the hospital, to improve patient safety and satisfaction (Demirhan and Pınar 2014). Tolerating normal diet well and becoming mobilized adequately by the patient, maintaining sufficient pain control and eliminating intravenous fluid dependency have been stated as discharge criteria (Ersoy and Gündoğdu 2007, Solak-Kabataş and Özbayır 2016).

In the literature, randomized controlled clinical and meta-analysis studies have reported that patients followed-up with the ERAS protocols have shortened total duration of hospital stay and fallen half complications (Adamina et al. 2011, Thorell et al. 2016, Yang et al. 2017, Li et al. 2018, Persico et al. 2018).

Follow-Up

Discharged patients should be called by phone 24-48 hours apart; their situation should be got to know. If there is no problem, the patient should be invited on the seventh to tenth days postoperatively to control the wound and remove the sutures (Demirhan and Pınar 2014). The ERAS protocol will come to the fruition via sharing and discussing the data. Thus the protocol will be developed and, if necessary, halting points will be changed (Ersoy and Gündoğdu 2007, Demirhan and Pınar 2014).

The Role of Nurses in Implemention of the Accelerated Healing Protocol after Surgery

Not only surgeons but also anesthesiologists, nurses, and physiotherapists actively participate in the protocol, which is different from traditional approaches, and even patients are included (Karen et al. 2015, Ljungqvist et al. 2017, Aksoy et al. 2018).

Since surgical nursing covers nursing care before, during and after the surgery, it continues from the admission of the patient to the discharge of the patient; nurses play active roles in every step of the surgical procedure (Demirhan and Pınar 2014, Hübner et al. 2015). The ERAS protocol is one of the current evidence-based approaches to which surgical nurses should adapt (Mitchell 2011, Karen et al. 2015, Çilingir and Candaş 2017). For this reason, surgical nurses should know their dependent and independent roles in the ERAS protocol, use these information in patient care, follow the current publications about the ERAS protocol and be open to develop. Evidence based nursing practices provide quality care in terms of bio-psychosocial health, improve the quality of care and its outcomes, create a difference in clinical practice and patient care outcomes, standardize care, minimize administration errors and increase nurse satisfaction (Karen et al. 2015, Aksoy et al. 2018).

Surgical nurses should continue patient education during the surgical procedure by especially identifying the problems that individuals may experience after discharge with good planning in the preoperative period. The fact that the ERAS is a multidisciplinary approach; the presence of surgical nurses at every step supports this consideration (Karen et al. 2015, Çilingir and Candaş 2017). In this approach, nurses have a key role in fulfilling the somatic and physiological needs of patients. The ERAS protocol is a model which consists of the applications involving nurses' dependent and independent roles. In the ERAS protocol, nurses have a role in psychological support especially for decreasing the stress of patients, restarting normal diet by patients, early mobilization and early rehabilitation of patients as soon as possible (Solak-Kabataş and Özbayır 2016). Nurses are also primarily responsible for the monitoring of the surgical procedure because nurses are health professionals in contact with patients for 24 hours therefore they also primarily responsible for the application and maintenance of the ERAS program (Hübner et al. 2015, Ljungqvist et al. 2017, Watson 2017). According to the literature, it has shown that the surgical nurses use their time more effectively and have lower workload as a result of the application of the ERAS protocol (Bianchini et al. 2014, Hübner et al. 2015, Pędziwiatr et al. 2015, Persico et al. 2018).

REFERENCES

Abdikarim, I., Cao, X. Y., Li, S. Z., Zhao, Y. Q, Taupyk, Y., Wang, Q. (2015). Enhanced recovery after surgery with laparoscopic radical gastrectomy for stomach carcinomas. *World J Gastroenterol, 21*, 13339–44.

Adamina, M., Kehlet, H., Tomlinson, G. A., Senagore, A. J., Delaney, C. P. (2011). Enhanced recovery pathways optimize health outcomes and resource utilization a meta-analysis of randomized controlled trials in colorectal surgery. *Surgery, 149*, 830-40.

Alfonsi, P., Slim, K., Chauvin, M., Mariani, P., Faucheron, J. L., Fletcher, D. (2014). French guidelines for enhanced recovery after elective colorectal surgery. *Journal of Visceral Surgery, 151*, 65-79.

Aksoy, A. & Vefikuluçay-Yılmaz, D. (2018). A new approach to evidence based practices in gynecological surgery: Eras protocol and nursing. *Turkiye Klinikleri J Nurs Sci, 10*, 49-58.

Bianchini, C., Pelucchi, S., Pastore, A., Feo, C. V., Ciorba, A., (2014). Enhanced recovery after surgery (ERAS) strategies: possible advantages also for head and neck surgery patients? *Eur Arch Otorhinolaryngol, 271*, 439-43.

Çam, R. (2015). *Taylor Klinik Hemşirelik Becerileri: Perioperatif Hemşirelik.* [Taylor Clinical Nursing Skills: Perioperative Nursing.] Ankara, Nobel Academic Publishing.

Çilingir, D. & Candaş, B., (2017). Enhanced Recovery After Surgery Protocol and Nurse's Role. *Journal of Anatolia Nursing and Health Sciences, 20*, 137-43.

Demirhan, İ. & Pınar, G., (2014). Enhanced Recovery After Surgery And Nursing Practices Enhanced Recovery And Nursing. *Yıldırım Beyazıt University Journal of Nursing, 2*, 43-53.

Gustafsson, U. O., Scott, M. J., Schwenk, W., Demartines, N., Roulin, D., Francis, N., McNaught, C.E., Macfie, J., Liberman, A. S., Soop, M., Hill, A., Kennedy, R. H., Lobo, D. N., Fearon, K., Ljungqvist, O., (2013). Guidelines for perioperative care in elective colonic surgery enhanced recovery after surgery (ERAS) society recommendations. *World J. Surg, 37*, 259–84.

Hübner, M., Addor, V., Slieker, J., Griesser, A.C., Lécureux, E., Blanc, C., Demartines, N., (2015). The impact of an enhanced recovery pathway on nursing workload: A retrospective cohort study. *International Journal of Surgery, 24*, 45-50.

Jimenez, R. S., Alvarez, A. B., Lopez, J. T., Jimenez, A. S., Conde, F. G., Saez, J. A. C. (2014). *Colorectal Cancer - Surgery, Diagnostics and Treatment: ERAS (Enhanced Recovery after Surgery) in colorectal surgery.*

Karen, M. B., Deborah S. K., Conor P. D., (2015). Successful Implementation of an enhanced recovery pathway: The nurse's role. *AORN J, 102*, 470-78.

Khan, S., Gatt, M., Horgan, A., Anderson, I., Macfie, J., (2009). *Guidelines for implementation of enhanced recovery protocols. Association of Surgeons of Great Britain and Ireland.*

Kehlet, H. & Wilmore, D. W., (2005). Fast-track surgery. *British Journal of Surgery, 92*, 3–4.

Li, M., Zhang, J., Gan, T. J., Qin, G., Wang, L., Zhu, M et al., (2018). Enhanced recovery after surgery pathway for patients undergoing cardiac surgery: a randomized clinical trial. *Eur J Cardiothorac Surg*, 1-7.

Ljungqvist, O., Scott, M. D., Fearon, K. C., (2017). Enhanced recovery after surgery a review. *JAMA Surg*. 152, 292-98.

Mehta, O. H. & Barclay, K. L., (2014). Perioperative hypothermia in patients undergoing major colorectal surgery. *Anz Journal of Surgery, 84, 550-55.*

Miles, V., Gonzalez, A., Amos, V., (2015). Aprepitant: A new modality for the prevention of postoperative nausea and vomiting: An evidence-based review. *Journal of Peri Anesthesia Nursing, 30*, 406-17.

Mitchell, M. (2011). The future of surgical nursing and enhanced recovery programmes. British *Journal of Nursing, 20*, 978-84.

Neville, A., Lee, L., Antonescu, I., Mayo, N. E., Vassiliou, M. C., Fried, G. M., Feldman, L. S., (2014). Systematic review of outcomes used to evaluate enhanced recovery after surgery. *Br J Surg, 101*, 159-70.

Oshvandi, K., Shiri, F. H., Fazel, M. R., Safari, M., Ravari, A., (2014). The effect of pre-warmed intravenous fluids on prevention of intraoperative hypothermia in cesarean section. Iranian *Journal of Nursing and Midwifery Research, 19*, 64-69.

Pędziwiatr, M., Kisialeuski, M., Wierdak, M., Stanek, M., Natkaniec, M., Matłok, M., Major, P., Małczak, P., Budzyński, A., (2015). Early implementation of enhanced recovery after surgery (ERAS®) protocol compliance improves outcomes: A prospective cohort study. *Int J Surg, 21*, 75-81.

Persico, M., Miller, D., Way, C., Williamson, M., O'Keefe, K., Strnatko, D., Wright, F., (2018). Implementation of enhanced recovery after surgery in a community hospital: an evidence-based approach. *J Perianesth Nurs*, 1-10.

Ronco, M., Iona, L., Fabbro, C., Bulfone, G., Palese, A., (2012). Patient education outcomes in surgery: a systematic review from 2004 to 2010. *Int J Evid Based Healthc, 10*, 309-23.

Solak-Kabataş, M. & Özbayır, T., (2016). Enhanced recovery after surgery (ERAS) protocols after colorectal surgery: A systematıc revıew. *Gümüşhane University Journal of Health Sciences, 5*, 120-32.

Thorell, A., MacCormick, A. D., Awad, S., Reynolds, N., Roulin, D., Demartines, N., Vignaud, M., Alvarez, A., Singh, P. M., Lobo, D. N., (2016). Guidelines for perioperative care in bariatric surgery: enhanced recovery after surgery (ERAS) society recommendations. *World J Surg, 40*, 2065-83.

Tunç-Tuna, P. & Kurşun, Ş., (2018). Enhanced recovery after surgery protocols and nursing care for colon surgery. *DEUHFED, 11*, 180-88.

The Turkish Anaesthesiology and Reanimation Society [TARS]. (2013). The prevention of inadvertent perioperative hypothermia *Turk J Anaesthesiol Reanim, 41*, 188-90.

Varadhan, K. K., Lobo, D. N., Ljungqvist, O., (2010). Enhanced recovery after surgery: the future of improving surgical care. *Crit Care Clin, 26*, 527-47.

Watson, D. J. (2017). The role of the nurse coordinator in the enhanced recovery after surgery program. *Nursing, 47*, 13-17.

Yang, L., Kaye, A. D., Venakatesh, A. G., Green, M. S., Asgarian, C. D., Luedi, M. M., Liu, H., (2017). Enhanced recovery after cardiac surgery: an update on clinical implications. *Int Anesthesiol Clin, 55*, 148-62.

In: Advances in Health and Natural Sciences ISBN: 978-1-53614-639-4
Editors: Burcu Yuksel et al.

Chapter 6

CURRENT APPROACHES TO PREVENT DIABETIC FOOT ULCERS

Ayfer Peker
Health Care Services (Podology), Kocaeli Üniversity, Kocaeli, Turkey

ABSTRACT

Diabetes mellitus (DM) is a common metabolic disorder that can cause various complications. According to the literature, the prevalence of diabetes is rapidly increasing both in Turkey and worldwide. DM is an important health issue due to the high costs of healthcare and serious complications that may affect patients in different ways. Diabetic foot ulcer is a major complication of DM. Although highly preventable, diabetic foot ulcer places a mental, physical, social, and economic burden on patients, and it significantly affects their quality of life. Besides longer bed occupancy and higher healthcare costs, diabetic foot ulcer also increases the risk of lower limb amputation and mortality. Almost 12-25% of diabetic patients will suffer from diabetic foot ulcer during their lifetime, and 10-15% of patients admitted with diabetic foot require surgical intervention. The amputation rate in diabetic foot disease throughout Europe was reported as 0.5-0.8%. In the United States, 85% of lower-extremity amputations are preceded by foot ulcers in patients with DM. Evidence-based guidelines for the management of diabetic foot problems have noted that diabetic foot ulcers can be prevented to a great extent through regular foot examinations and proper therapeutic measures. The treatment and care of diabetics require a multidisciplinary approach because of the extensive nature of the problem. Regular diabetic neuropathy assessment, vascular assessment, joint and muscle examination, and walking analysis are thus recommended. Care of the pressure points on the plantar aspect of the foot is also essential to prevent diabetic foot ulcers. The main objective of a healthcare team in the management of a diabetic patient is to achieve metabolic control with an emphasis on health education, and to improve survival and quality of life.

Keywords: diabetic foot ulcers, prevention, podology, podiatry, walking analysis, personal shoes and insoles

INTRODUCTION

Diabetes mellitus (DM) is one of the most common chronic diseases. It is expensive to treat and its complications have a major impact on health expenses. DM was predicted to affect about 171 million people worldwide in 2000; this was predicted to rise to more than 340 million in 2030 (Wild et al., 2004). DM leads to multifaceted complications. Diabetic foot ulceration is one of the common complications related to the disease, and it can result in lower limb amputations (Boulton and Vileikyte, 2000; Reiber, 2001; Dang and Boulton, 2003; Pinzur et al., 2005).

Foot and lower extremity complications have become an increasingly significant public health concern in people with diabetes in both developed and developing countries. Foot problems are the most important and expensive complications of DM. These complications, such as neuropathy, ischemia and diabetic foot ulcers, typically lead to infection and lower extremity amputation. However, it is possible to prevent amputation in diabetic patients. An approach that involves prevention, patient education, staff education, a multidisciplinary diabetic foot team, and foot screening can reduce amputation rates by 49–85% (Stephanie et al., 2007).

The etiological factors for the development of diabetic foot ulcers can be determined in a clinical environment using inexpensive, everyday equipment designed for foot examination. Preventing diabetic foot ulceration is possible through foot screenings which analyze the protective sensation via the Semmes-Weinstein monofilament, as well as vascular (Doppler ultrasound and ankle-brachial blood pressure), orthopedic, dermatological, biomechanical (plantar foot pressure) issues and include a physical examination with a brief history. These measurements, and the findings of the physical examinations, help health providers to categorize patients based on risk and to decide what approach to use. Regular foot care, patient education, glycemic control and quitting smoking are other effective clinical interventions (Stephanie et al., 2007).

ETIOLOGY AND PATHOPHYSIOLOGY OF DIABETIC FOOT ULCERS

The etiology of diabetic foot ulcers usually has many constituents. The most frequent underlying etiologic factors of diabetic foot ulcers are peripheral neuropathy, trauma, foot deformity and high plantar pressures. Peripheral neuropathy plays a central role in most patients: up to 50% of diabetic patients have diabetic peripheral neuropathy and are at risk of developing diabetic foot ulcers. Neuropathy causes an insensitive and deformed foot, often with problems in walking. Walking barefoot and/or in ill-fitting shoes can induce a chronic ulcer in neuropathic people, as can acute trauma. Thickened skin (callus) grows as a result of the abnormal biomechanical loading of foot due to a foot deformity, limited joint mobility, poor-fitting shoes or a loss of sensation. A subcutaneous hemorrhage forms after this abnormal biomechanical loading of the foot. Peripheral vascular problems may result in an ischemic foot ulcer caused by a minor trauma. Symptoms may be absent in patients with neuropathy, ischemia or neuroischemic foot (Frykberg et al., 2000; Boyko et al., 1999,

Pecoraro et al., 1990; Armstrong and Lavery 1998; IWGDF Guidelines, 2011).

Most of the risk factors for foot ulceration are also predisposing risk factors for amputation, because foot ulcers are the prime reason for amputation (Pecoraro et al., 1990; Armstrong and Lavery, 1998). Ulcerations can result in limb loss for two reasons. They can to lead poor wound healing and to progressive tissue necrosis in the presence of critical ischemia (Armstrong and Lipsky, 2004).

PREVENTION

The prevention of diabetic foot ulcers in diabetic patients is a major clinical goal of healthcare professionals, as foot problems can result in lower extremity amputations. Studies of the causes of amputation show that ulcers lead to 84% of diabetes-related amputations. (Pecoraro et al., 1990). Prophylactic foot care has been shown to be connected with decreased morbidity, lower health costs and a reduced risk of amputation (Pinzur et al., 2005). Approaches that involve the classification of risk factors, as well as patient and family education and podiatric care, have been proven to be cost-effective and to reduce costs (Moreland et al., 2004; Singh et al., 2005; Ragnarson et al., 2004). Proven strategies for prevention of ulceration involve having a *multidisciplinary team approach to foot care, patient screening and follow-up, professional foot care for pre-ulcerative signs (debridement of hyperkeratosis, callus, corn and nails), patient education and therapeutic shoes* (Pecoraro et al., 1990; Dargis et. al., 1999; Apelqvist and Agardh, 1992).

Multidisciplinary Foot Care

Diabetic foot is a frequent problem and leads to high costs. A multidisciplinary foot approach to diabetic foot ulcerations has been shown to reduce the cost of diabetic foot treatment. Studies from around the world show a larger per capita expense for diabetics in general compared to other patients. This high cost is more evident when associated with diabetic foot problems, and is correlated to the length of hospitalization. Multidisciplinary team work with diabetic foot problems has been shown to decrease ulcer and amputation rates, and reduce the costs of the treatment itself (Matricali et al., 2007).

A team of dedicated health care professionals is essential for preventing lower limb amputation. Those responsible for foot care may include a diabetologist, vascular surgeon, podiatrist, orthopedic surgeon, pedorthist, general surgeon, and plastic surgeon. New studies have shown that podiatrists and vascular surgeons are the primary members of the diabetic foot care team. A podiatrist conducts regular screening and offers professional foot care; when a vascular problem appears a vascular surgeon carries out revascularization and after the operation the podiatrist continues to provide foot care (Lee et al., 2010). An ostomy nurse, specialist diabetes nurse, dietician and physical therapist can also participate in the team if required (Lee et al., 2010).

A prospective study noted fewer recurring ulcerations in diabetic foot ulcer patients after the formation of multidisciplinary diabetic foot care teams (Oyibo et al., 2001). Studies from Asia have shown that teamwork and the frequent implementation of preventive diabetic foot protocols strongly decreased the incidence of amputation (Burns et al., 2003). It has also been proven that the use of multidisciplinary diabetic foot teams including podiatrists reduced amputations by 34% (Connor and Oz, 2004).

Patient Screening and Follow-Up

Patient screening and follow-up means to evaluate the patient's actual risk of developing a diabetic foot ulcer or undergoing amputation using the risk classification system and to carry out reassessments at regular intervals. If there is no peripheral neuropathy, the patient should have a foot examination once a year; if the patient has peripheral neuropathy, they should have a foot examination once every 6 months; if the patient has peripheral neuropathy with peripheral artery disease and/or a foot deformity, they should have a foot examination once every 3-6 months; if the patient has peripheral neuropathy and a history of foot ulcer or a lower-extremity amputation, they should have foot examination once every 1-3 months.(NICE, 2015). A foot examination can be repeated very often (every 1–2 weeks etc.) for high risk patients if there is urgent concern.

Professional Foot Care for Pre-Ulcerative Signs (Debridement of Callus and Nails)

Professional foot care for pre-ulcerative signs means the treatment of any pre-ulcerative sign on the foot. This includes:

- Treatment of calluses;
- Treating blisters and making them safe;
- Treatment of ingrowing or hypertrophic/thick nails;
- Treatment of hemorrhages;
- Treatment protocols for fungal infections.

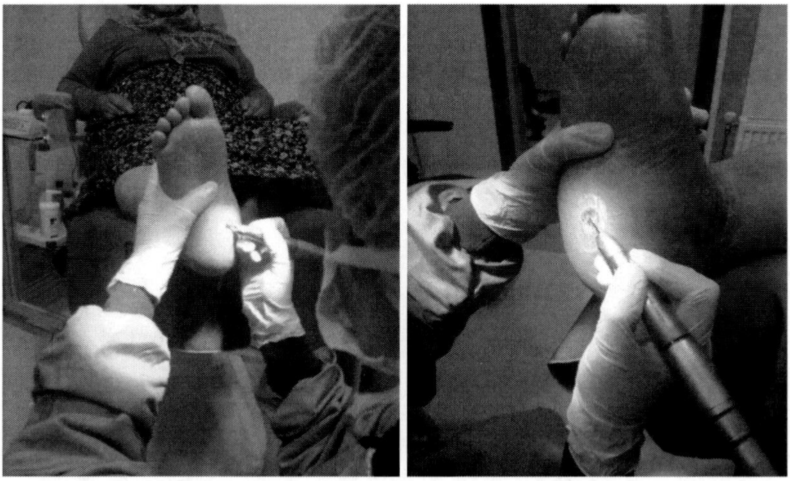

Figure 4. Removing calluses.

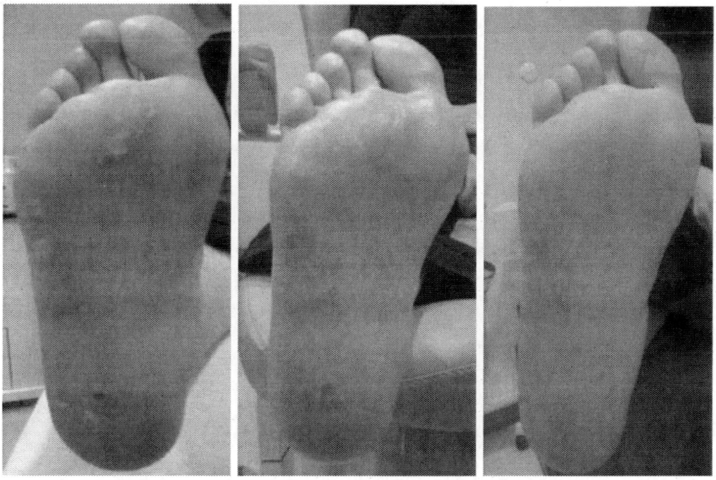

Figure 5. Callus before and after treatment and at 18 months' follow-up.

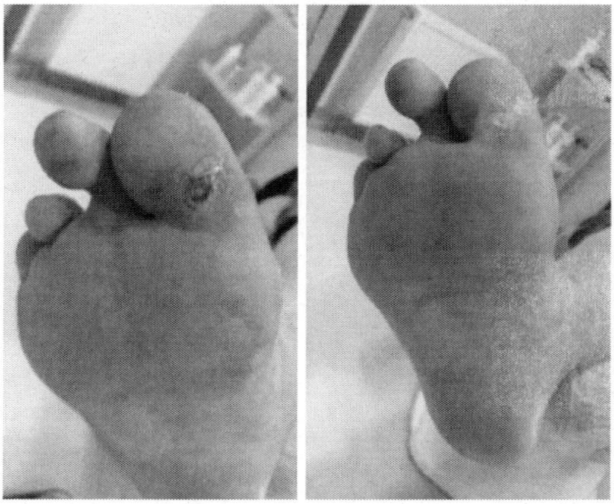

Figure 6. Treatment of pre-ulcerative lesion.

The usefulness of the treatment of pre-ulcerative signs for preventing foot ulcers has not yet been clearly studied. However, removing calluses has been shown to be beneficial as it decreases plantar pressure which is a significant risk factor for ulcer formation (NICE, 2014; IWGDF, 2015).

Some images of our patients receiving foot care treatment are provided in Figures 4-10.

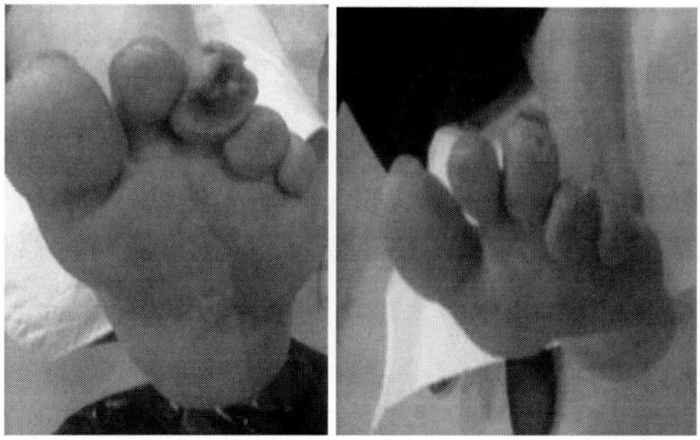

Figure 7. Treatment of a neuropathic ulcer.

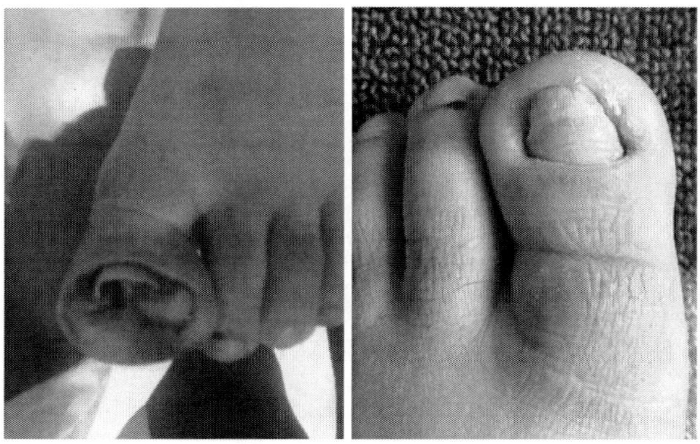

Figure 8. Treatment of an ingrown nail with nail brace.

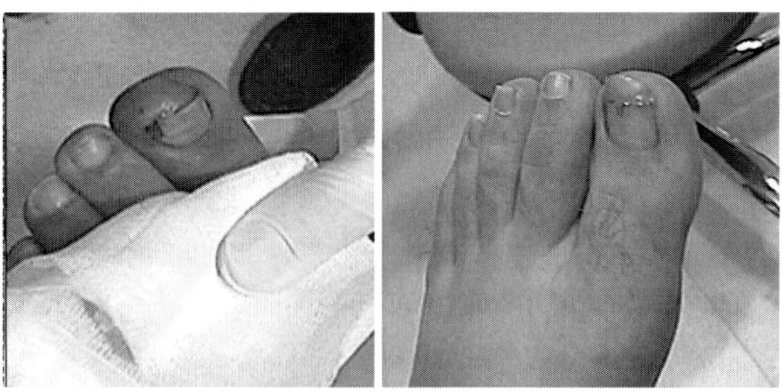

Figure 9. Treatment of an ingrown nail with nail brace.

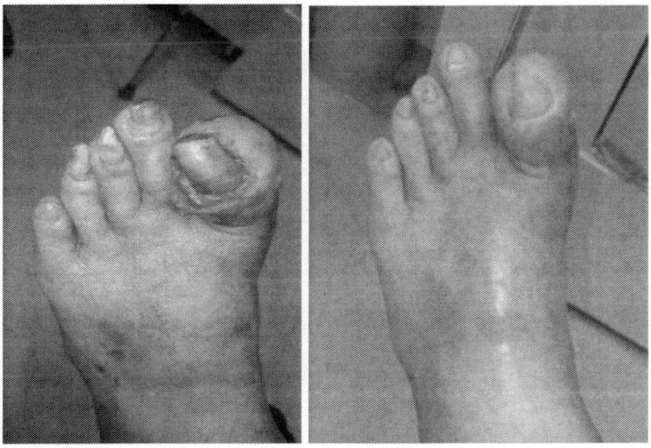

Figure 10. Treatment of hypertrophic nail.

One randomized controlled study (RCT), one non-RCT and three non-controlled studies showed a dramatically lower rate of recurring foot ulceration in patients who received holistic foot examinations and care compared with those who did not (Dargis et al., 1999; Plank et al., 2003; Hamonet et al., 2010; Marcinia and Chantelau, 1998).

Patient Self-Management Education

Education of the patient and their family should aim to provide the correct information about foot care and how to practice it. It is very important to support the patient in following the advice given. The education should involve how to self-manage their condition, what to do in foot care emergencies, recommendations about shoes, the patient's risk of developing a foot ulcer, and information about DM and blood glucose control. Patient education should be repeated regularly (NICE, 2015).

A diabetic person should practice the following self-management activities in an organized manner:

- Inspection of the feet and the inside of shoes every day;

- Washing the feet every day (drying them carefully especially between the toes);
- Refraining from using chemicals and plasters to treat callus and corns;
- Using balsam oil on the skin;
- Cutting nails straight;
- Monitoring the temperature of the skin on the feet to avoid the recurrence of plantar foot ulcerations (NICE, 2015; IWGDF, 2015).

Therapeutic Footwear

Foot ulcerations/amputations are connected to poorly fitting shoes in diabetic patients so off-loading techniques and different shoes are commonly using for the prevention and treatment of foot problems. Wearing incorrect shoes both indoors and outdoors is a major cause of foot ulcers. Shoes should be adapted to changing biomechanics and to foot deformities. Diabetic people with normal protective sensation can choose their shoes themselves. In people with peripheral neuropathy/ischemia and foot deformities extra attention must be paid when fitting footwear (Pecoraro et al., 1990; Dargis et al., 1999; Apelqvist and Agardh, 1992; Armstrong and Lipsky, 2004).

The patient should be directed to use special diabetic shoes, slippers, socks and insoles at home and when outside. To prevent a recurrent non-plantar ulcer, patients need to use therapeutic shoes, custom-made insoles or a toe orthosis (Chantelau, 1993; IWGDF, 2015).

Patients should use custom-made shoes and insoles to treat a plantar foot ulcer as standard or conventional therapeutic shoes are inadequate for treatment. Traditional half shoes which are designed for off-loading of the forefoot and which only support the heel and midfoot, are unsuitable because of the risk of a midfoot fracture (IWGDF, 2015).

Casual footwear, shoe modifications and toe orthoses can be used to treat and produce off-loading for non-ischemic, non-plantar or under control infection ulcers. Patients should wear well-fitting footwear for a recurrent non-plantar ulcer. Custom-made insoles and shoes, therapeutic shoes and a toe orthosis are required if a foot deformity or a pre-ulcerative sign is present (IWGDF, 2015).

REFERENCES

Apelqvist J. & Agardh C. D. (1992) The association between clinical risk factors and outcome of diabetic foot ulcers. *Diabetes Res Clin Pract,* 18, 43–53.

Armstrong D. G. & Lavery L. A. (1998) Diabetic foot ulcers: prevention, diagnosis and classification. *Am Fam Physician*, 57, 1325–32.

Armstrong D. G. & Lipsky B. A. (2004). Advances in the treatment of diabetic foot infections. *Diabetes Technol Ther*, 6, 167–77.

Boulton A. J. & Vileikyte L. (2000). The diabetic foot: the scope of the problem. *J Fam Pract*, 49(11), 3–8.

Boyko E. J. &Ahroni J. H. &Stensel V. & Forsberg R. C. & Davignon D. R. & Smith D. G.(1999). A prospective study of risk factors for diabetic foot ulcer: the Seattle Diabetic Foot Study. *Diabetes Care*, 22, 1036–42.

Burns P. &Gough. S. & Bradbury A. W. (2003) Management of peripheral arterial disease in primary care. *BMJ,* 326, 584-88.

Chantelau E., Breuer U., Leisch A. C, Tanudjaja T. & Reuter M. (1993) Outpatient treatment of unilateral diabetic foot ulcers with 'half shoes.' *Diabetic medicine: a journal of the British Diabetic Association*, 10(3), 267-70.

Connor H. & Mahdi O. Z. (2004) Repetitive ulceration in neuropathic patients. *Diabetes Metab Res Rev*, 20 (1), 23–28.

Dang C. N. & Boulton A. J. (2003). Changing perspectives in diabetic foot ulcer management. *Int J Low Extrem Wounds*, 2, 4–12.

Dargis V., Pantelejeva O., Jonushaite A., Vileikyte L. & Boulton A. J. (1999) Benefits of a multidisciplinary approach in the management of recurrent diabetic foot ulceration in Lithuania: a prospective study. *Diabetes Care*, 22, 1428–1431.

Frykberg R. G. & Armstrong D. G. & Giurini J. & Edwards A. & Kravette M. & Kravitz S. & et al. (2000). Diabetic foot disorders: a clinical practice guideline. American College of Foot and Ankle Surgeons. *J Foot Ankle Surg*, 39, S1–60.

Hamonet J. & Verdie-Kessler C. & Daviet J. C. & Denes E. & Salle J. Y. et al. (2010), Evaluation of a multidisciplinary consultation of diabetic foot. *Annals of Physical and Rehabilitation Medicine*, 53, 306-318.

I WGDF *Guidance on use of interventions to enhance the healing of chronic ulcers of the foot in diabetes* (2015) Available from: URL: http://iwgdf.org/guidelines/development-of-the-iwgdf-guidance-documents-2015/.

Lee C. R. & George A. & Joseph C. & Lawrence B. & Harkless, J. L. & Mills S. & Armstrong D. G. (2010)Toe and flow: Essential components and structure of the amputation prevention team. *Journal Of Vascular Surgery,* 52,12.

Marcinia M. & Chantelau E.(1998) Qualified podiatry for rehabilitation of patients with diabetic foot syndrome. A cohort study. *Diabetes und Stoffwechsel*, 7, 81-85.

Matricali G. A. & Dereymaeker G. & Muls E. & Flour M. & Mathieu C. W. (2007) *Diabetes/Metabolism Research and Reviews*, 23, 339 – 347.

Moreland M. E. & Kilbourne A. M. et al. (2004) Diabetes preventive care and non-traumatic lower extremity amputation rates. *J Health Qual,* 26, 12–7.

NICE guideline. *Diabetic foot problems: prevention and management* (2015). Available from: URL:https://www.nice.org.uk/guidance/ng19.

NICE guideline. *Diabetic foot problems: prevention and management* (2014) Available from: URL: https://www.nice.org.uk/guidance/cg179.

Oyibo S. O. & Jude E. B. & Tarawneh I. & Nguyen H. C. & Harkless L. B. & Boulton A. J. (2001) A comparison of two diabetic foot ulcer classification systems: the Wagner and the University of Texas wound classification systems. *Diabetes Care*, 24, 84–88.

Pecoraro R. E. & Reiber G. E. & Burgess E. M. (1990) Pathways to diabetic limb amputation. Basis for prevention. *Diabetes Care*, 13, 513–21.

Pinzur M. S. & Slovenkai M. P. et al. (2005). Guidelines for diabetic foot care: recommendations endorsed by the Diabetes Committee of the American Orthopaedic Foot and Ankle Society. *Foot Ankle Int,* 26, 113–19.

Plank J. & Haas W. & Rakovac I. & Görzer E. &Sommer R. & Siebenhofer A. (2003) Evaluation of the impact of chiropodist care in the secondary prevention of foot ulcerations in diabetic subjects. *Diabetes Care*, 26, 1691-1695.

Ragnarson T. G. & Apelqvist J. (2004) Health-economic consequences of diabetic foot lesions. *Clin Infect Dis*., 39, 132–9.

Reiber G. E. & Ledoux W. R. (2001). Epidemiology of foot ulcers and amputations in the diabetic foot. In: Editor Herman W. & Kinmonth A. L. &Wareham N. & Williams R. *The Evidence Base for Diabetes Care*, USA; Black Well; 2010, 403-415.

Singh N. & Armstrong D. G. et al. (2005) Preventing foot ulcers in patients with diabetes. *JAMA*, 293, 217–28.

Stephanie C. W. & Vickie R. D. & James S. W. & Armstrong D. G. (2007) Foot ulcers in the diabetic patient, prevention and treatment. *Vasc Health Risk Manag*, 3(1), 65–76.

Wild S. & Roglic G. & Green A. & Sicree R. & King H. (2000). Global prevalence of diabetes; estimates for the year 2000 and projections for 2030. *Diabetes Care, 27*, 1047–1053.

In: Advances in Health and Natural Sciences ISBN: 978-1-53614-639-4
Editors: Burcu Yuksel et al.

Chapter 7

A Review of Studies on the Application of Coumarin and Its Derivatives in Plants

Burcu Yuksel**
Kocaeli Vocational School of Health Services,
Kocaeli University, Kocaeli, Turkey

Abstract

This chapter summarizes the use of coumarin, a secondary metabolite, and its derivatives in different aspects of plant growth and formation. Coumarin and its derivatives have broad biochemical and pharmacological activities. They also have allelopatic effects on plants. In this chapter, summaries and review of previous research are presented by pointing out significant contributions and effects of coumarin and its derivatives in plant growth and inhibition. Nowadays, the active use of coumarin and its derivatives, which are used as defense elements in several areas such as agriculture, food, cosmetics and medicine, necessitates the clarification of action mechanisms of coumarin. In order to understand these compound's allelopathic and toxic effects completely,

* Corresponding Author Email: burcu.yuksel@yahoo.com.

it is essential to carry out cytological, genetic, biochemical and physiological tests altogether. In the studies on plants, it is very important to understand the defense mechanisms in plant species that make them resistant to stress conditions in order to minimize yield loss in terms of nutrition and agricultural economics.

Keywords: allelochemical, coumarin, cytotoxicity, genotoxicity, plants

INTRODUCTION

With their broad range and biomass, distribution of plants that constitute the first step of food chain is determined by the biotic (plants that belong to the same species, plants that belong to the different species, animals, bacteria and fungi) and abiotic (temperature, geographical formations, soil composition, water and solar rays) factors. One of these factors, secondary metabolites, are the components that are excreted by the plants. These components affect the choice of the plants' habitats to a large extent. These components' mechanism of action can produce inhibitor effects to a certain extent and may generate physiological, cytotoxic or genetic changes in the organism. Hereditary variations that emerge when these components compete with the organisms' DNA may, in turn, cause decrease in their population or extinction. Also, these components may affect product quantity and quality in a negative way by generating physiological and biochemical damage. The plants appeal to their molecular defense mechanisms in order to minimize or prevent negative effects completely. These components may generate adverse outcomes in the organisms when they are consumed. Coumarin is one of the highly studied secondary metabolites.

Coumarin was purified in its free form for the first time in 1820 by Vogel from the dried sweet-smelling seeds of *Dipteryx odarata* (*Coumarouna odarata*) tree called 'tonka bean,' which grows in South America from the Fabaceae family. In plants, coumarin and coumarin derivatives are commonly found on their own or compositely. Because of their various biological activities, these kinds of compounds are natural compounds that have gained significance in recent years. Coumarin's chemical structure is first explained by Strecker (1867) and Fitting (1870) according to Coffey (1977). Today, apart from the tonka bean, coumarin derivatives are isolated from 600 kinds of plants. Coumarin is found, especially, in Orchidaceae, Leguminoseae, Rutaceae, Apiaceae and Lamiaceae families' plants' fruits, leaves, roots, stems and shell sections at a certain rate (Boga, 2005).

Crystal-like, white and solid coumarins are composed of aromatic rings that are fused with condense lactone ring. 1,2-benzopiran, 2*H*-1-benzopyran-2-one, *cis*-o-coumaric acid lactone and coumarinic anhydride are the other names that are used commonly. Its molecular formula is $C_9H_6O_2$ and its molecular weight is 146,14 g/mol. While coumarin's melting point is 68-70°C, its boiling point is 297-299°C. Coumarins can be soluble in ethanol, chloroform, diethyl ether and oil, but it is slightly soluble in water. While first class coumarins are isolated from Indian beans, coumarins, which are used for commercially, are synthesized by salicylaldehyde (Yuce, 2006).

Herbal coumarin derivatives have broad biochemical and pharmacological activities. Cinnamic acid and its derivatives, which are the pioneers of coumarin and its derivatives, are the substantial compounds of sweeteners, perfumes, synthetic dye and pharmaceutics (Bickers et al. 2005). Because of their significant bioactivity power, new types of coumarin derivatives, which have efficient pharmacological value, have been synthesized in recent years. Antibiotic, anti-coagulant, anti-cancer, anti-inflammatory and bacteriostatic effects that are found in coumarin and its derivatives, make them a significant matter in biology and medicine (El-Khatib et al. 2007). They also have antioxidant, vasodilator, anti-inflammatory, immune stimulant, antiallergic and antiviral effects. At the same time, the studies conducted on humans and animals have indicated that coumarins decrease the risk of catching cardiovascular and cancer diseases (Cirico et al. 2006). Although it was used in food industry because of its redolence for a period of time, the studies that were conducted on animals revealed its toxicity and its use was abandoned in this field (Guenther, 1975). These phenolic compounds' toxic aspects are revealed by different researchers. According to the previous research, phenolic compounds like tannic acid, coumarin, vanillin, caffeine and cinnamic aldehyde had mutagenic effects on mammalian cells (Sanyal et al. 1997). High doses of phenolic compounds' genotoxic effects occurred by the mechanisms like DNA intercalation, DNA topoisomerase II damage, free radical production and key enzyme inhibition (Stopper et al. 2005).

When the topic is approached from plant stress physiology perspective, the agents that are applied on plants generate various changes on plant metabolism. In order to determine physiological and biochemical changes, one has to have knowledge about the transmitter and the target plant. At the same time, knowing these agents' structure and mechanism of action eases the description of morphological, cytological and genetic changes that they cause. For instance, abnormal rhizogenesis indicates that inhibitor material prevents the microtubule formation. Lateral rhizogenesis or damage of primary root indicates that hormonal balance is destroyed. The difference between the length of the root and stem indicates that plant growing is affected at the primary level. These effects depend on the impact of the compound's mechanism of action and biologic effect, which have extracted an allelochemical effect, as well as the target plant's physical and biochemical tolerance status (Erez, 2009). As a result of the secondary metabolites' or allelopathic compounds' interaction with the target plant, temporary pores may open on the cell membrane or they may cause membrane disintegration. Thus, by increasing cell membrane permeability, they may make them open to exogenic materials (Wink, 2010). Also, these compounds can easily react with proteins and toxic materials, which have a significant role in metabolic and structural events, gene regulation and interaction. As a result of this interaction, protein conformation may easily become active or inactive. Secondary metabolites perform this effect by connecting proteins covalently, especially with their own reactive functional groups (Wink & Latzbruning, 1995). Moreover, allelopathic agents, which contain phenol, may decrease, increase or cap off the plant hormones' physiological activities. This situation inhibits plants' normal physiological development process. In general, phenolic compounds' effect on germination and early seedling development may be activating or inhibiting depending on the compound that is used, concentration and target plant (Reigosa et al. 1999).

PREVIOUS STUDIES ON THE EFFECTS OF COUMARIN ON PLANTS

Coumarin's significant effects such as acting like an antioxidant, an enzyme inhibitor and in repressing toxicants are examined in plant biochemistry and physiology (Yuksel, 2017; Abenavoli et al. 2003; Saleh et al. 2015). Coumarins in plants facilitate or inhibit plant development depending on their types and concentrations. A plant affects call differentiation and auxin metabolism by acting like development promoter (Abenavoli et al. 2003; Demiralay, 2005; Yuce, 2006).

Coumarin's influence on *Vicia faba* L was examined by considering some parameters like germination and growth in the study of Saleh et al. in 2015. To determine coumarin's influence on germination, the *Vicia faba* L seeds were treated with coumarin in different concentrations. It is stated that coumarin influenced the plant more in growth phase, significantly enhanced phytohormone rates as well as primary and secondary metabolites. Results of Saleh et al.'s study indicated that coumarin might influence plant's growth and physiology as an active growth substance and it might have an interaction with phytohormone metabolism.

In Razavi's (2011) study, it is stated that some coumarin derivatives like imperatorin and psoralen may be used for producing eco-friendly pesticide chemicals. Mata et al. (1998) used *Amaranthus hypochondriacus,* which is an imperatorin of a natural coumarin derivative on *Echinochloa crusgalli, Lactuca sativa* and *Lycopersicum esculentum.* They examined the effects of coumarin on radicle growth and germination. Based on their findings, they reported that, imperatorin inhibited ATP synthesis and phosphorylation mechanisms, and acted like an allelopathic agent on these plants.

Wu et al. (2016) applied 40 µg $mL^{-1'}$ coumarin on Italian rye flower (*Lolium multiflorum Lam.*), common knotgrass (*Polygonum aviculare*), red clover (*Trifolium pratense*), veronica (*Veronica persica*), *Poa pratensis* L, *Chenopodium album* and *Plantago asiatica* plants and found out that coumarin, in these plants, significantly inhibited the seed germination and seedling growth. It is stated that important inhibition effects denoted both seed germination and seedling growth on all coumarin applied test plants in 80 µg mL^{-1}. They also stated that coumarin completely hindered Italian rye flower's and red clover plant's seed germination and seedling growth. Results of their study revealed that coumarin was the dominant allelochemical substance of yellow sweet oat and this substance had a powerful inhibition effect on seed germination and seedling growth on many weeds. According to the results of this study, coumarin may be used as a natural herbicide.

Razavi et al. (2010) conducted a study on *Zosima absinthifolia* plant which belongs to the Apiaceae family. This plant is used as food aroma and food product in Iran. In this study, imperatorin, which is one of the plant seeds known as furanocoumarin, 7-preniloksi coumarin, and auraptene were purified. Results of their study revealed that imperatorin displayed a fungi toxic effect against *Sclerotinia sclerotiorum,* which is a common plant pathogen. It is stated that compounds also had phytotoxic effects and they played an allelochemical role.

Findings of Shettel and Balke's (1983) study revealed that coumarin derivative umbelliferon inhibited the growth of *Zea mays* L., *Glycine max* L, *Abutilon tbeophrasti* Medic., *Amaranthus retroflexus* L., and *Panicum miliaceum* L. plants. In the study of Kupidlowska et al. (1994), the effects of coumarin and xanthotoxin substances on *Allium cepa* plant were examined. It is reported that these substances retarded mitosis in *Allium cepa* root tips meristematic cells, decreased oxygen uptake by its meristematic cells, created structural changes in mitochondrial matrix and caused changes in organelles' formation and physiology. The results of third study revealed that coumarin increased succinate dehydrogenase enzyme production more than xanthotoxin, and energy metabolism was affected as a result of the subsequent changes. Similar to the findings of Razavi et al. (2010), they found that coumarin acted like allelochemicals.

In another study, Brooker et al. (2007) examined coumarin's six different derivatives in terms of their antifungal activity against pathogen fungi on various soil plants. They were specifically interested in *Macrophomina phaseolina* and *Pythium* spp., pathogens amongst these fungi. Their results supported previous studies about natural insecticide development with the help of coumarin and its derivatives.

Al-Barwani et al.'s (2004) study found that coumarin derivative furanocoumarins were dominant antifungal compound on *Conium makulatumun* plant. Along with these compounds' absence in healthy plant structurally, they increased 103 times more in plants that got stressed compared to the control group. The findings of their study suggested that these compounds might be considered as stress compounds because they increased as a result of chemical stress.

There were some other studies on coumarin, which examined its insecticidal effects. For example, Olson and Roseland (1991) and Stevewnson et al. (2003) found that coumarin and its derivatives had insecticidal effects. In the light of the findings of these studies, it is stated that working on coumarin may bring a new approach to the development of genetic plants that produce coumarin by using genetical engineering and to the control of insect herbivores.

In another study Ahrabi et al. (2011) investigated the biochemical and physiologic reactions of para-hydroxybenzoic acid and coumarin, which are the secondary metabolites from phenolic compounds that affect plant growth and development on canola. Chlorophyll, carotenoid and anthocyanin ingredients were identified and catalase activities were measured. Moreover, lipid peroxidation was measured with thiobarbituric acid. The results of the study revealed that seedling growth decreased significantly. It is stated that chlorophyll ingredient decreased with the increase in phenolic compounds' concentrations however carotenoids and antioxidants like anthocyanin that had low molecular weight decreased significantly. It is reported that antioxidant enzyme guaiacol peroxidase activity increased along with the increase in phenolic compound concentration but catalase activity did not change significantly. In spite of the increase in the antioxidant capacity of canola seedlings, it is explained that lipid peroxidation increased in the presence of these phenolic compounds. The results revealed that some plants' allelopathic potential was related with its phenolic compounds and allelochemical research may cause development of herbicides that had less negative effect on environment and ecosystem. It is reported that these phenolic acids' presence may decrease canola yield and it is prominent to destroy these weeds for the production of canola in terms of an allelopathic angle.

In their study, Abenavoli et al. (2006) investigated allelochemical coumarin on *Triticum turgidum sperması durum* Desf. plant. Findings of their study revealed that coumarin inhibited the seed germination in high concentrations. More specifically, coumarin blocked the water intake, lessened the electrolyte storage capacity and O_2 consumption during early germination stage and delayed the activation of peroxidases. It also increased superoxide dismutase activity, decreased some enzymes' (APX, CAT) activities and suppressed some molecular chaperons' transcriptions. In another study, Aliotta et al. (1992) investigated coumarin's effects on radish. The results of the study revealed that coumarin was an effective inhibitor and it inhibited the root growth of carboxylic group in the side chain.

Zhou et al. (2013) investigated the effects of coumarin application on the growth of *Medicago Sativa*, net nitrate intake, nitrate (NO_3^-) and ammonium (NH_4^+) concentration, the amount of soluble protein and the effects on nitrate reductase, glutamine synthase and glutamate dehydrogenase activities. They found out that coumarin decreased the fresh weight production and root/body development rate at shoots and roots in high concentrations. Net nitrate intake was stimulated significantly at the seedlings that were exposed to coumarin in low concentrations but it diminished with the increase in coumarin amount. Their results also revealed that while low coumarin application increased NO_3^- ingredient prominently, it decreased in high coumarin application. NH_4^+ rate, on the other hand, decreased in low coumarin concentration, but increased in high concentration. It is reported that nitrate reductase activities were suppressed in high coumarin concentrations and glutamine synthase rates decreased at first but then increased. It is also stated that glutamate dehydrogenase activities were precluded at roots in the highest coumarin concentrations. According to these results, it is stated that coumarin in low concentrations stimulated net nitrate intake and N metabolism at roots but in high concentrations it affected the basic enzymes of nitrogen metabolism in clover.

Based on the results of their study, Li and Gao (2011) stated that a coumarin derivative 4-metilumbelliferon (4-MU) inhibited seed germination by decreasing primary root growth in *Arabidopsis thaliana*. It is also found that 4-MU contributed to the hypocotyl formation and auxin distribution. In another study, Abenavoli et al. (2003) applied coumarin (1,2 benzopyrone) on *Daucus carota* L. Cv. Saint Valery. As a part of their study, they generated plant culture in order to investigate the effects on cell growth and nitrate intake and the use of ammonium and carbohydrates on this plant. While coumarin application caused cell growth inhibition, free amino acids, ammonium rates and glutamine synthetase; glutamate dehydrogenase, glukoz-6-fosfat dehydrogenase and phosphoenolpyruvate carboxylase enzyme activities increased in cells. It is stated that malate dehydrogenase, on the other hand, was inhibited. These effects were interpreted from the point of protein catabolism's stimulation and/or intervention with the protein biosynthesis which was inducted by coumarin. Similar to other studies discussed in this review, in this study, it is stated that allelochemical compound coumarin might act as an inhibitor of cell cycle and/or as a senescence-promoting substance.

In another study, Abenavoli et al. (2004) investigated the effects of coumarin on corn seedlings (*Zea mays* L. cv. Cecilia) when the parameters like the length, diameter of different root types and branching density were taken into consideration. The amount of applied coumarin inhibited the root length. It is also found that especially 1 and 0.25 mM coumarin concentration decreased radical, seminal and nodal root lengths, respectively. In the light of the results that they found, it is stated that nodal > seminal > radical plant roots were more sensitive to coumarin.

Another study that investigated the allelochemical effects of coumarin was Lupini et al. (2010). It focused on the effects of coumarin on root elongation speed at three different maize primer root zones, transition zone and two non-growing zones. ATPase activity, pH changes and cell membrane potentials at root zones were evaluated. According to the results of the study, it is stated that coumarin increased root elongation speed at root transition zone thus, it caused an increase in ATPase activity. It is reported that cell membrane permeability changed by causing proton amount change.

In a different study, Al-Wakeel et al. (2013) applied three different concentrations of coumarin and salicylic acid against charcoal rot disease caused by *Macrophomina phaseolina* (Tassi) Goidanich on sunflower (*Helianthus annuus* L.). It is stated that the preparation of sunflower seed with 0.3 mM coumarin or 1.0 mM salicylic acid reduced the degree of disease and provided approximately 50% protection on the length of root lesions at greenhouse conditions and more than 80% decrease. It is reported that both coumarin and salicylic acid applications caused accumulation of soluble sugars and phytoalexins, at the same time they also stimulated β-1,3-glucanase and chitinase activities.

Pergo et al.'s (2008) study focused on nine different plants against *Bidens pilosa* L in terms of phytotoxicity. *Bidens pilosa* L grows in Brazil and it is responsible for yield loss at some products. In their study, while coumarin treatment inhibited *Bidens pilosa* L development, it did not affect the development of surrounding useful plants (*Senna obtusifolia* L., *Euphorbia heterophylla* L. ve *Ipomoea grandifolia* L.). Among the results of the study, it is stated that coumarin acted like a cytostatic agent which retarded germination.

Saleh and Madany's (2015) study investigated the effects of coumarin in increasing endurance against salinity stress of the plant. They covered wheat germs with coumarin (50ppm) and then grew them under different NaCl levels (50, 100, 150 mM) during two weeks. Because of the accumulation of osmolytes like soluble sugars and proline with coumarin pretreatment, it enhanced the growth of wheat germs under salinity compared to the seedlings that coumarin was not applied. Also, coumarin treatment significantly increased K^+/Na^+ rates at plants' seedlings' shoots that were under salinity stress. In addition to this, it is stated that at roots, this rate increased in the plans that have no salinity stress. It is found that phenolic materials and flavonoids accumulated with coumarin treated Phenylalanine ammoniac lyase (PAL). Salt stress significantly stimulated peroxidase (POD) activity at the seedlings that were pretreated with coumarin. Additionally, the effects of coumarin on total antioxidant capacity was obtained only with the highest NaCl doze (150 mM).

Saleh and Kebeish (2017) demonstrated that coumarin and its derivatives had different physiologic effects on tissue and organ levels. In their study, coumarin's effect on redox homeostasis of wheat germs at aleurone layers was investigated. The results revealed that both coumarin and GA3 implementations induced α-amylase production but inhibited superoxide dismutase, catalase and ascorbate peroxidase activities. Antioxidant enzymes' downregulation by coumarin and GA3 caused both H_2O_2 and significant accumulation of malondialdehyde. The results showed that coumarin might ruin redox balance at aleurone layers by inactivating enzymatic antioxidant system. It is stated that, coumarin may show gibberellin-like activity. In another study, Saleh and Abu El-Soud (2015) also found that coumarin had a gibberellin-like effect and it was effective in germination and sapling formation.

In the investigation of new plant killer compounds that cover weeds, Araniti et al. (2015) examined the effects of three synthetic coumarin derivatives on two harmful weeds, *Amaranthus retroflexus* and *Echinochloa crus'* germination and rhizogenesis. All of the coumarin derivatives that they synthesized affected both weed germination and rhizogenesis. Findings of their study revealed that coumarin and its derivatives might have promising herbicide activity potential.

Niro et al. (2016) define coumarins as having allelopathic characteristics. In their study, researchers investigated the effects of different doses of coumarin on a weed (wild oat), crop (wheat) and microbial community in natural soil conditions. Findings of their study reaffirmed that coumarin was a natural herbicide that had a prominent potential in sustainable agriculture. In another study, Grana et al. (2017) evaluated the mechanism of action and phytotoxic potential of scopoletin, which is a natural compound that belongs to the coumarin group. Results of their study demonstrated that scopoletin caused mitochondrial membrane depolarization and rhizogenesis that stimulated cell death. With the analysis conducted, they observed cell and tissue anomalies at roots, such as cell wall deformations, multinucleate cells, abnormal nuclei and tissue disorder.

Lupini et al. (2014) also defined coumarin as an active allelopathic compound that played a key role in plant-plant interaction and communication. The interaction between coumarin and auxin was examined at *Arabidopsis thaliana* (Col-0) on root system's development. It is stated that coumarin strongly affected *Arabidopsis* seedling's primary root extension and root development. It is also found that especially 10^{-4} M coumarin significantly inhibited main root extension by increasing lateral root numbers and root bristle length. Also, coumarin addition increased the negative effects of TIBA and NPA which were the two auxin transport inhibitors that caused lateral root formation's inhibition. These results revealed that there was a functional interaction between coumarin and auxin in the root development of *A. thaliana*.

Yuksel and Aksoy (2017) examined the cytological effects of Coumarin (2H-1-benzopyran-2-one) on *Lens Culinaris Medic* in terms of cell response, mitotic index, mitotic abnormalities and chromosome aberrations. Germination percentages of *L. culinaris* roots decreased with increasing coumarin concentrations. Cytological observations demonstrated that the mitotic frequency in root meristematic cells decreased and that abnormality frequency also decreased in parallel to the increase in concentrations of coumarin. The results indicated that this active ingredient had the ability to cause a reduction in the seed germination percentage in the number of different phases of mitosis and increased chromosomal aberration of *L. culinaris*.

Conclusion

As I have tried to present in this chapter, coumarin and its derivatives have a wide range of applications in different plants for different purposes. Nowadays, the active use of coumarin and its derivatives, which are used as defense elements, in several areas such as agriculture, food, cosmetics and medicine, necessitates the clarification of action mechanisms of coumarin. In order to understand these compound's allelopathic and toxic effects completely, it is essential to carry out cytological, genetic, biochemical and physiological tests together. With the studies that are conducted on plants, it is highly important to understand the defense mechanisms in plant species that are resistant to stress conditions in order to minimize yield loss in terms of nutrition and agricultural economics. For this purpose, most of the studies have focused on the production of plant species that are resistant to stress condirions in recent years. These studies are are also offering solutions to prevent nutrition problems that are likely to emerge in the near future.

References

Abenavoli, M. R., Sorgonà, A., Sidari, M., Badiani, M., & Fuggi, A. (2003). Coumarin inhibits the growth of carrot (Daucus carota L. Cv. Saint Valery) cells in suspension culture. *Journal of Plant Physiology*, *160*(3), 227-237.

Abenavoli, M. R., Sorgonà, A., Albano, S., & Cacco, G. (2004). Coumarin differentially affects the morphology of different root types of maize seedlings. *Journal of Chemical Ecology*, *30*(9), 1871-1883.

Abenavoli, M. R., Cacco, G., Sorgona, A., Marabottini, R., Paolacci, A. R., Ciaffi, M., & Badiani, M. (2006). The inhibitory effects of coumarin on the germination of durum wheat (*Triticum turgidum ssp.* durum, cv. Simeto) seeds. *Journal of Chemical Ecology*, *32*(2), 489.

Ahrabi, F., Entesharisup, S., Moradshahisup, A. (2011) Allelopathic potential of para-hydroxybenzoic acid and coumarin on canola: Talaieh cultivar. *Journal of Medicinal Plants Research*, 5(20), 5104-5109.

Al-Barwani, F. M., Eltayeb, E. A. (2004). Antifungal compounds from induced *Conium maculatum L.* plants, *Biochemical Systematics and Ecology*, 32(12), 1097-1108.

Al-Wakeel, S., Gabr, M., Abu-El-Soud, W., Saleh, A. (2013). Coumarin and salicylic acid activate resistance to *Macrophomina phaseolina* in Helianthus annuus. *Acta Agronomica Hungarica*, 61(1), 23-35.

Aliotta, G., Cafiero, G., Fiorentino, A., Strumia, S. (1993) Inhibition of radish germination and root growth by coumarin and phenylpropanoids. *Journal of Chemical Ecology*, 19(2), 175-183.

Araniti, F., Mancuso, R., Lupini, A., Giofrè, S. V., Sunseri, F., Gabriele, B., Abenavoli, M. R. (2015) Phytotoxic potential and biological activity of three synthetic coumarin derivatives as new natural-like herbicides. *Molecules, 20*(10), 17883-17902.

Bickers, D., Calow, P., Greim, H., Hanifin, J. M., Rogers, A. E., Saurat, J. H., Sipes, I. G., Smith, R. L., Tagami, H. (2005) A toxicologic and dermatologic assessment of cinnamyl alcohol, cinnamaldehyde and cinnamic acid when used as fragrance ingredients. *Food and Chemical Toxicology*, 43, 799-836.

Boğa, M. (2005). *Kumarin Türevi Yeni 14:4 Taç Eter Bileşiklerinin Sentezi ve Yapılarının Aydınlatılması, Yüksek Lisans Tezi.* Marmara Üniversitesi, Fen Bilimleri Enstitüsü, İstanbul, Türkiye. [*The synthesis of novel 14:4 crown ethers derivatives of coumarin compound and characterization of their structures*, Marmara University, Institute of Applied Sciences, Istanbul, Turkey].

Brooker, N. L., Kuzimichev, Y., Laas, J., Pavlis, R. (2006) Evaluation of coumarin derivatives as anti-fungal agents against soil-borne fungal pathogens. *Communications in agricultural and applied biological sciences*, 72(4), 785-793.

Cirico, T. L., Omaye, S. T. (2006) Additive or synergetic effects of phenolic compounds on human low density lipoprotein oxidation, *Food and Chemical Toxicology*, 44, 510-516.

Coffey S., *Rood's Chemistry of Carbon Compounds,* Elsevier, New York, Vol. IV(E), 2nd Ed, 1977.

Demiralay, N. (2005). Bazı Kumarin Bileşiklerinin Asitlik Sabitlerinin Tayini ve Fe(II), Cu(II), Ni(II) *Metal İyonları ile Kompleksleşmelerinin İncelenmesi*, Yüksek Lisans Tezi, Marmara Üniversitesi, Fen Bilimleri Enstitüsü, İstanbul, Türkiye. [*Determination of acidity constants of some coumarin compounds and investigation of complexation of these compounds with Fe(II), Cu(II), and Ni(II) metal ions*, Master's thesis, Marmara University, Institute of Applied Sciences, Istanbul, Turkey.

El-Khatib, R. M., Nassr, L. A. E. (2007). Reactivity trends of the base hydrolysis of coumarin and thiocoumarin binary aqueous methanol mixtures at different temperature. *Spectrochimica Acta*, 67: 643-648.

Erez M. E. (2009). Lepidium draba L., Acroptilon repens (L.) DC., Thymus kotchyanus Boiss Hohen. var. kotchyanus, Inula peacockiana (Aitch. &Hemsl.) Koravin, Salvia kronenburgei *Rech. f.* ve *Phlomis armeniaca* Wild. bitkilerinin allelopatik etkilerinin araştırılması. Doktora Tezi, Yüzüncüyıl Üniversitesi, Fen Bilimleri Enstitüsü, Van, Türkiye. [Investigation of allelopatic potential of *Lepidium draba L., Acroptilon repens (L.) DC., Thymus kotchyanus Boiss & Hohen. var. kotchyanus, Inula peacockiana (Aitch. & Hemsl.) Koravin, Salvia kronenburgei Rech. f. and Phlomis armeniaca Wild.* plants, Unpublished doctoral dissertation, Yuzuncuyil University, Institute of Applied Sciences, Van, Turkey.

Graña, E., Costas-Gil, A., Longueira, S., Celeiro, M., Teijeira, M., Reigosa, M. J., & Sánchez-Moreiras, A. M. (2017). Auxin-like effects of the natural coumarin scopoletin on Arabidopsis cell structure and morphology. *Journal of plant physiology*, *218*, 45-55.

Guenther E., *The Essential Oils, Vol. II*, D. Von Nostrand Ltd., New York, 1975.

Kupidlowska, E., Dobrzynska, K., Parys, E., Zobel, A. M. (1994). Effect of coumarin and xanthotoxin on mitochondrial structure, oxygen uptake, and succinate dehydrogenase activity in onion root cells. *Journal of chemical ecology*, 20(10), 2471-2480.

Li, X., & Gao, M. J. 2011. Modulation of root branching by a coumarin derivative. *Plant signaling & behavior*, 6(11), 1654-1655.

Lupini, A., Sorgonà, A., Miller, A. J., & Abenavoli, M. R. (2010). Short-term effects of coumarin along the maize primary root axis. *Plant signaling & behavior*, 5(11), 1395-1400.

Lupini, A., Araniti, F., Sunseri, F., Abenavoli, M. R. (2014). Coumarin interacts with auxin polar transport to modify root system architecture in *Arabidopsis thaliana*. *Plant growth regulation,* 74(1), 23-31.

Mata, R., Macías, M. L., Rojas, I. S., Lotina-Hennsen, B., Toscano, R. A., Anaya, A. L. (1998). Phytotoxic compounds from Esenbeckia yaxhoob. *Phytochemistry*, 49(2), 441-449.

Niro, E., Marzaioli, R., De Crescenzo, S., D'Abrosca, B., Castaldi, S., Esposito, A., Rutigliano, F. A. (2016). Effects of the allelochemical coumarin on plants and soil microbial community. *Soil Biology and Biochemistry*, *95*, 30-39.

Olson, M. M., & Roseland C. (1991). Roseland, Induction of the coumarins scopoletin and ayapin in sunflower by insect-feeding stress and effects of coumarins on the feeding of sunflower bettle (*Coleontara chrysomelidae*). *Environ. Entomol*., 20: 1166-1172.

Pergo, É. M., Abrahim, D., Da Silva, P. C. S., Kern, K. A., Da Silva, L. J., Voll, E., Ishii-Iwamoto, E. L. (2008). *Bidens pilosa* L. exhibits high sensitivity to coumarin in comparison with three other weed species. *Journal of chemical ecology*, 34(4), 499.

Razavi, S. M., Imanzadeh, G. H., Davari, M. (2010) Coumarins from Zosima absinthifolia seeds, with allelopatic effects. *EurAsia. J. Biosci*, 4, 17-22.

Razavi SM, Plant Coumarins as allelopathic agents. *International Journal of Biological Chemistry*, 2011, 5: 86-90.

Reigosa, M. J., Souto X. C., Gonz. L. (1999) Effect of phenolic compounds on the germination of six weeds species. *Plant Growth Regul*, 28:83–88.

Saleh, A. M., El-Soud, W. A. (2015) Evidence for "gibberellin-like" activity of coumarin. *South African Journal of Botany*, 100, 51-57.

Saleh, A. M., Madany, M. M., González, L. (2015) The effect of coumarin application on early growth and some physiological parameters in Faba Bean (*Vicia faba* L.). *Journal of Plant Growth Regulation*, 34(2), 233-241.

Saleh, A. M., & Madany, M. M. Y. (2015). Coumarin pretreatment alleviates salinity stress in wheat seedlings. *Plant Physiology and Biochemistry*, 88, 27-35.

Saleh, A. M., Kebeish, R. (2017). Coumarin impairs redox homeostasis in wheat aleurone layers. *Journal of Plant Research*, 1-7.

Sanyal, R., Darroudi, F., Parzefall, W., Nagao, M., Knasmüller, S. (1997). Inhibition of the genotoxic effects of heterocyclic amines in human derived hepatoma cells by dietary bioantimutagens. *Mutagenesis*, 12, 297-303.

Shettel, N. L., & Balke, N. E. (1983). Plant growth response to several allelopathic chemicals. *Weed Science*, 31(3), 293-298.

Stevewnson, P., M. S. J. Simmonds, M. A. Yule, N. C. Veitch, G. C. Kite, D. Irwin M., Legg, (2003). Insect antifeedent furanocoumarins from *Tetradium daniellii*. *Phytochemistry*, 63: 42-46.

Stopper, H., Schmitt, E., Kobras, K. (2005). Genotoxicity of phytoestrogens. *Mutation Research*, 574, 139-155.

Wink, M., & Latzbruning, B., (1995). Allelopathic properties of alkaloids and other natural-products - possible modes of action. *Allelopathy*, 582 117-126.

Wink M., Functions and biotechnology of plant secondary metabolites. *Blackwell Pub*, 2010.

Wu, C. X., Zhao, G. Q., Liu, D. L., Liu, S. J., Gun, X. X., Tang, Q. (2016). Discovery and weed inhibition effects of coumarin as the predominant allelochemical of yellow sweetclover (*Melilotus officinalis). International Journal of Agriculture & Biology*, 18(1).

Yuce B., (2006). Hesperidin, Rutin ve 7, 8-Dihidroksi-3-(4-Metilfenil) *Kumarin bileşiklerinin lipit düşürücü ve antioksidan etkilerinin araştırılması*, Yüksek Lisans Tezi, Marmara Üniversitesi Fen Bilimleri Enstitüsü, İstanbul, Türkiye. [*The investigation of lipid lowering and antoxidant effects of hesperidin, rutin and 7,8-dihydroxy-3-(4-*

methylphenyl) coumarin, Master's thesis, Marmara University, Institute of Applied Sciences, Istanbul, Turkey].

Yuksel, B. (2017). *Determination of cytotoxic, biochemical and genotoxic effects of coumarin on lentil* (*Lens Culinaris* Medic). Unpublished Doctoral Dissertation. Institute of Applied Sciences, Kocaeli University, Turkey.

Yuksel, B., & Aksoy, O. (2017). Cytological effects of coumarin on the mitosis of *Lens Culinaris* Medik. *Fresenius Environmental Bulletin*, *26*(11), 6400-6407.

In: Advances in Health and Natural Sciences ISBN: 978-1-53614-639-4
Editors: Burcu Yuksel et al.

Chapter 8

Awareness of Brucellosis: A Worldwide Zoonosis

Mustafa Sencer Karagul*
Kartepe Vocational School of Equine Science,
Kocaeli University, Kocaeli, Turkey

Abstract

Brucellosis, which is one of the most common zoonotic diseases, causes abortion, infertility and reduced milk production in animals. Brucellosis in humans, on the other hand, is defined as an acute or sub-acute febrile illness including an intermittent or remittent fever. Because of brucellosis, many countries run the risk of suffering from economic and health problems like losing husbandry and infection risks. Brucella, which dates back to the 5th plague of Egypt around 1600 BC in history and which was named after the research of Sir David Bruce, is an important disease and brucella genus has 11 species. The aim of this chapter is to define this disease and its species and biovars, to explain its historical background, and to reveal the ways it affects animals and humans.

Keywords: biovar, *Brucella melitensis*, brucellosis, Malta fever, zoonosis

---

* Corresponding Author Email: msencerk@hotmail.com.

INTRODUCTION

Brucellosis is one of the most common zoonotic diseases caused by *Brucella (B.)* spp., which causes important health and economic problems (Godfroid et al., 2005; Songer & Post, 2012; Yumuk& O'Callaghan 2012). This disease, which has subacut or chronic forms, affects many different animal species (Corbel, 2006). It causes both economic and health problems through the loss of husbandry and it brings about risks due to the infections appearing via dairy products (Pappas et al., 2006a; Seleem et al., 2010). The distribution of host includes humans, wild and domestic animals such as sheep, goat, cattle, camel, pig, boar, rabbit, dog, marine mammals, fox, and baboon (Corbel, 2006; Godfroid et al., 2005; OIE, 2016; Songer &Post, 2012). It is defined as a chronic contagious disease causing necrotic inflammatory infections and complications such as abortion, infertility, arthritis, orchitis and mastitis in susceptible hosts (Alton et al., 1988; Aydin, 1997). Moreover, World Health Organization (WHO) laboratory biosecurity manual describes *Brucella* organisms as belonging to risk group 3 microorganisms (OIE, 2009a; OIE, 2012; WHO, 2004).

As *Brucella* species can be transmitted via aerosol way, it is classified as a potential bioterror agent as well. It is also stated that there are no vaccines for human beings (Godfroid et al., 2005; Songer & Post, 2012). Because of its significance as a potential agent of bioterrorism, the pathogen is on the category B biodefense research list of the Centers for Disease Control and Prevention (CDC) and the National Institute of Allergy and Infectious Diseases (NIAID). The non-stratified lists of potential biological weapons of other organizations, such as the WHO, the North Atlantic Treaty Organization (NATO), and the Biological and Toxin Weapons Convention (BTWC) also include brucellosis (Pappas et al., 2006a). When it is used as a weapon for agroterrorism, it might ruin the economy (DelVecchio & Mujer, 2013).

Brucellosis is known as '*undulant fever*', "*Mediterranean fever*" or "*Malta fever*" (Corbel, 2006; Godfroid et al., 2011; Matthew & Leland, 2004) and it is transmitted through infected animals or their products

(Corbel, 2006). However, Mediterranean fever, which is synonymous with brucellosis, should not be confused with familial Mediterranean fever. (Mattew & Leland, 2004). Brucellosis, especially caused by *Brucella melitensis*, remains one of the most common zoonotic diseases worldwide with more than 500,000 human cases reported annually (Bosilkovski et al., 2009; Pappas et al., 2006b; Seleem et al., 2010).

Brucellosis in cattle is usually caused by biovars of *Brucella abortus*. In some countries, especially in southern Europe and western Asia, where cattle are *kept* together with sheep or goats, *B. melitensis* might cause this infection (OIE, 2012). Brucellosis in cattle is known as Bang's disease and epizootic abortion and contagious abortion are the terms that also used in veterinary medicine. (Matthew & Leland, 2004). Brucellosis in sheep and goats (excluding *Brucella ovis* infection) is primarily caused by one of the three biovars of B. *melitensis*. (OIE, 2009a). *Brucella ovis* causes a disease unique to sheep and it is one of the most common causes of epididymitis in rams and a rare cause of infertility and abortion in ewes and neonatal mortality in lambs (OIE, 2015a). Sporadic infections caused by *B. abortus* or *B. suis*, which are rare cases, have been observed in sheep and goats. On the other hand, it is possible to observe *Brucella melitensis* infection in domestic and wild susceptible species when these species are kept together with sheep and goats in enzootic areas (OIE, 2009a). Out of 5 biovars of *B. suis*, biovars 1, 2 and 3 are responsible for the infection in pigs (OIE, 2009b). Canine brucellosis, which is a zoonatic disease and caused by *Brucella canis*, is an important cause of reproductive failure, particularly in kennels (CFSPH, 2012).

THE HISTORY OF THE DISEASE

It is possible to trace the roots of this disease in the 5th plague of Egypt around 1600 BC. When ancient Egyptian bones dating to around 750 BC were examined, some evidence of sacroiliitis and other osteoarticular lesions, common complications of brucellosis was found out (Pappas, 2006a; Seleem et al., 2010). In addition, a researcher named L. Capasso found the bone lesions which were typical of brucellosis in the skeletal remains of people who died during the first volcanic eruption of Mount Vesuvius. He also analyzed a buried carbonized cheese through electron microscopy and his analysis revealed that there were coco-like forms which were morphologically consistent with *Brucella* spp. (D'anastasio et al., 2011; Godfroid et al., 2005; Seleem et al., 2008). Since people domesticated animals, brucellosis has affected human populations. There is evidence that Hippocrates might have mentioned it in his book titled 'On Epidemics' and it was a common disease in the ancient world (Matthew & Leland, 2004).

The disease, which is known as brucellosis, was first found out in Malta in 1850s. It was the British medical officers serving on the island who spotted the disease after the Crimean War (Wyatt, 2013). In 1859, Marston made one of the earliest descriptions of the disease in 1859 except for the biblical references (Nicoletti, 2002). He also described typhoid fever and differentiated it from '*Undulant fever'*, what is known as brucellosis (Nicoletti, 2002; Wyatt, 2013). The research of Sir David Bruce with the military and civilian population in Malta started the modern understanding with the help of the identification of a micrococcus from a patient who died of undulant fever in 1887 and it was named 'Micrococcus melitensis' (Godfroid et al., 2005; Matthew & Leland, 2004; Poester et. al. 2013; Wyatt, 2013). In 1905, again in Malta, Themistocles Zammit isolated it from goat's milk accidentally and thanks to this, he was able to illustrate the zoonotic nature of the disease. Before that, it was considered to be a vector-borne disease (Nicoletti, 2002; Seleem et al., 2010). This bacterium was renamed '*Brucella melitensis*' in Sir David Bruce's honor by Meyer and Shaw in 1920 (Poester et al., 2013).

Afterwards, L.F. Bernhard Bang, who was a Danish vet, mentioned an organism that he named *Bacillus abortus* which caused contagious abort cases in cattle, which was later found to be associated with Bruce's organism (Matthew & Leland, 2004; Nicoletti, 2002; Seleem et al., 2010) . The name of the disease was changed from 'Bovine infectious abortion to Bang's disease in 1930 (Nicoletti, 2002). An American scientist, Alice Evans, who carried out important work on pathogenic bacteria in dairy products identified once again the relationship between Bang's disease and Malta Fever and she renamed the genus *Brucella* to honour David Bruce (Parascandola, 1998; Seleem et al., 2010).

Brucellosis in pigs caused by *Brucella suis* was first reported in 1909 by Huntyra in Hungary (Ozcanarslan, 2011). Soon after that, in 1914, Jacob Traum isolated it from aborted swine fetuses in Indiana, USA. (Nicoletti, 2002; Ozcanarslan, 2011; Poester et al., 2013). *B. ovis*, which causes epididymitis in rams, was first isolated in New Zealand by McFarland and his colleagues. In 1953, Buddle ve Boyes stated that this agent was a variant of *B. melitensis*. This agent was named as *B. ovis* in 1956. (Nicoletti, 2002; Ozcanarslan, 2011; Poester et al., 2013). *B. neotomae* was first isolated in Utah, USA in 1957 from the wood rat Neotama lepida by Stoenner ve Lackman. This organism was defined as a new *Brucella* species on the basis of its behavior on differential dye media, CO_2 requirements, and H_2S production (Cameron & Meyer, 1958). *B. canis* was discovered in beagle dogs in the USA in 1966 and named as a distinct species by Carmichael in 1968 (Thimm, 1982). It is the germ of contagious abortion of kennel dogs and stray dogs as well. In 1994, British researchers isolated an unknown *Brucella* organism from stranded harbor seals (Phoca vitulina), harbor porpoises (Phocoena phocoena) and common dolphins (Delphinus delphis) on the coast of Scotland, and American researchers isolated it from an aborted foetus of a captive bottlenose dolphin (Tursiops truncatus) in California (Godfroid et al., 2005). The distinguishing metabolic profiles, dye and phage sensitivities of these isolates were found to be the same and then, they were labelled as *B. maris*. (Ozcanarslan, 2011). Later on, *B. maris* was categorized as *B. ceti*, which affects porpoises and dolphins, and *B. pinnipedialis*, which affects mainly seals. (Pappas, 2010). Previously, brucellosis was not recognized, not even suspected in marine mammals; however, as marine mammal *brucella* strains have been isolated, the ecologic range of the genus and its scope as a zoonosis have been extended. (Godfroid et al., 2005).

In Turkey, the first brucellosis case was identified by Husamettin Kural and Mahmut Sabit Akalin in 1915 in a soldier who was being treated in Istanbul Kuleli Military Hospital. The first isolation in cattle was made in 1931 by Berke and in sheep by Koylubag and Aktan in 1944 in Bandirma Merinos Farm (Aydın, 1997; Celebi, 2003; Cevik, 2001; Golem, 1949; Iyisan et al., 2000). Golem revealed the existence of brucellosis in humans and animals through serological methods (Aydın, 1997; Iyisan et al., 2000). In Turkey, the first documented case of brucellosis, which was confirmed by a laboratory, was in 1915 despite the fact that it is believed that Florence Nightingale got this disease during the Crimean War in Turkey (Young, 1995; Yumuk & O'Callaghan, 2012).

Brucella Species and Biovars

The microorganisms belonging to *Brucella* genus are not mobile; they are sporeless and gram-negative and in the form of small cocobacili (Aydin, 1997; OIE, 2012; Thimm, 1982). They are bacteria-like rods and they are 0.5 - 1.5μ in length and 0.5 - 0.7 μ in breadth (Alton et al., 1988; Corbel & Banai, 2005; OIE, 2012; Thimm, 1982). They grow in the presence of oxygen at an optimal temperature of 37°C (within the range of 20 - 40°C) and an optimal pH of 6.6-7.4.(Corbel & Brinley Morgan, 1982; Thimm, 1982; WHO, 1986). Another characteristic of *Brucella* spp. is that even though they are defined as immobile, they possess all the genes required to form a functional flagellum except the chemotactic system (Seleem et al., 2008; 2010).

In general, the morphology of *Brucella* strains is constant and pleomorphic forms are rare, except in old cultures growing under adverse conditions. Just because they are small and coccoide, they can be confused with cocci (Alton et al., 1988; Corbel & Banai, 2005). When serum or blood is added to the media, they foster the growth of *Brucella* spp. and a 5-10% CO2 atmosphere is a necessity for some strains of *B. abortus* (biovars 1 - 4 and 9) (Aydin, 1997; Perez-Sancho et al., 2015). They produce oxidase, catalase, nitrate reductase, urease; they do not produce indole; they are not hemolytic, do not melt gelatin, and also Methyl Red (MR) and Voges-Proskauer (VP) test results are negative. Most of them, except for *B. ovis*, use glucose as the source of energy (Songer & Post,

2012). When *Brucella* microorganisms are listed as taxonomies, the *Brucella* genus is defined as belonging to the family *Brucellaceae* within the order Rhizobiales of the class *Alphaproteobacteria* (Godfroid et al., 2011; Songer & Post, 2012). They belong to the alpha-2 subcategory of the *Proteobacteria*, together with *Ochrobactrum*, *Rhizobium*, *Rhodobacter*, Agrobacterium, Bartonella, and Rickettsia (Seleem et al., 2010). Molecular studies have shown that a non-pathogenic soil organism was possibly the predecessor of *Brucella* as true pathogens (Bardenstein et al., 2010). *Brucella* bacteria are genetically related to some plant pathogens and symbionts of the genera *Agrobacterium* and *Rhizobium*, in addition to animal pathogens (*Bartonella*) and opportunistic or soil bacteria (*Ochrobactrum*) (OIE, 2012). The closest phylogenetic neighbor of the genus *Brucella* is the genus O*rchrobactrum anthropic* as a saprophyte, which rarely infects animals (Godfroid et al., 2011). It is also claimed that *Brucella* organism is a species of *Orchrobactrum* (Pappas, 2010).

After *Brucella* bacteria have been observed in different animal species and different environments, *Brucella* genus is claimed to include 11 *Brucella* species (OIE, 2016). B. *abortus*, *B. melitensis*, *B. suis*, *B. ovis*, *B. neotomae* ve *B. canis* are considered to be classical species. This classification is based on pathogenic differences and host preferences (Altoparlak, 2003; ICSP, 2014; Vicente et al., 2014), specific phages and lysis sensitivity, their oxidation ways in various carbohydrate and amino acid subtrats (Gorvela & Moreno, 2002; Yumuk & O'Callaghan, 2012) and their other phenotypic characters. Out of 6 classical species, 5 of them lead to infection on one or more animal hosts and cause clinical symptoms. 4 of them; namely, *B. melitensis*, *B. suis*, *B. abortus* and *B. canis* (listed according to their pathogenicity) cause infection on humans (Corbel, 2006; Matthew & Leland, 2004; Seleem et al., 2010).

Biotyping is considered as a perfect way of finding and clarifying epidemiological data (Thimm, 1982). It is possible to divide *B. melitensis*, *B. abortus* ve *B. suis* into different biotypes based on the analysis of their requirements of CO_2, their production of H_2S, their growth on cultural media which contain dyestuffs, and their reaction to monospecific antisera (Cloeckaert et al., 2002; OIE, 2012; Thimm, 1982). *B. pinipedialis* and *B.*

ceti, which are hosted by marine mammals, are different from other *Brucella* species phenotypically with regards to their metabolic activity (Jahans et al., 1997; Seleem et al., 2010). The types which have recently been recognized as related to marine animals can also cause human disease (Corbel, 2006; Seleem et al., 2010). Another new species that has been recognized in central Europe in common voles and red foxes is *B. microti*. It is also considered to be a persistent soil contaminant. The agent has not been linked with human disease but if it has an anthropozoonotic risk, a particular target group would be hunters (Pappas, 2010). Hosts preferred mainly by *Brucella* species and their pathogenicity levels are shown in Table 1 (Godfroid et al., 2011; OIE, 2016).

Table 1. Brucella species, preferred hosts and pathogenicity for humans

Species	Biotype	Preferred hosts	Pathogenicity for humans
B. melitensis	1 - 3	Goat, sheep	High
B. abortus	1 - 6,9	cattle	High
B. suis	1 - 3	pig	High
	2	wild boar, hare	None[a]
	4	Reindeer, caribou	High
	5	Rodents	High
B. neotomae	-	Desert wood rat (Neotomalepida Thomas)	High
B. ovis	-	Ram	None
B. canis	-	Dog	Moderate
B. ceti	-	Cateceans	Unknown[b]
B. pinnipedialis	-	Pinnipeds	Unknown[b]
B. microti	-	Soil, vole, fox	Unknown
B. inopinata	-	Unknown	High
B. papionis	-	Unknown	Unknown

a. One case of *B. suis* biovar 2 infection in an immuno-compromised hunter has been described in France.

b. One human laboratory contamination has been described in the UK. Two naturally acquired cases have been described, although the source of infection could not be traced back to marine mammals.

Two new strains, in addition to the already recognized *Brucella* strains, have each been isolated from a single human case (Pappas, 2010). The first one (BO1T) has been labelled as *Brucella inopinata*. The name *'inopinata'* comes from Latin, which means '*unexpected*'. It was isolated from the breast implant wound fluid and blood of a female patient, who was 71 years old and who had the clinical symptoms of brucellosis. (Scholz et al., 2010). The second strain (BO2 strain) was isolated from the lung biopsy of a 52-year-old patient in Australia, who had a history of chronic destructive pneumonia. The analysis of the isolate has shown similar characteristics with *B. inopinata* (Tiller et al., 2010).

Strains are either smooth or rough, which express smooth lipopolysaccharide (S-LPS) and contain O-chain, or rough lipopolysaccaride (R-LPS), and which lack O-chain, as major surface antigen, respectively (Cloeckaert et al., 2002). S-LPS is the most significant antigenic form in the infections caused by smooth *Brucella* organisms and it serologically belongs to immunodominant epitopes, O-polysaccaride (O-PS) section. LPS layer of rough mutant *Brucella* bacteria lacks O-PS and they just carry coreolysaccaride-Lipid A part (Cloeckaert et al., 2002; Godfroid et al., 2011). LPS is a very important form for the structural and functional integrity of the outer membrane of the gram-negative bacteria. When compared to the classical enterobacterial LPS, the ones belonging to *Brucella* are a lot less active and less toxic than Escherichia coli (Seleem et al., 2008).

B. ovis ve *B. canis* are two *Brucella* species which are currently known as being naturally in the rough phase. *B. ovis* has similarities with other *Brucella* species in that they are alike in terms of their morphology, their staining properties and their cultural characteristics; however, the reactions of *B. ovis* to the oxidase and urease tests are negative. *B. ovis* colonies are not haemolytic. They are circular and convex and they have edges, which are unbroken. Moreover, they are always rough when they are examined by oblique illumination and test positive in the acriflavine test (Alton et al., 1988; OIE, 2015a). *B. ovis* requires an incubating atmosphere which has 5

- 10% CO_2 for growth. In addition, it is lysed by Rough/Canis phage (Alton et al., 1988; OIE, 2015a).

Finally, a new species, *B. papionis*, has been added to the *Brucella* genus (OIE, 2016). The phenotypic and genotypic analysis of 2 isolates obtained from stillborn baboon offsprings has shown that these isolates belong to *Brucella* genus. Detailed analysis of the strains has illustrated that these two strains have a new lineage within the genus *Brucella* (Whatmore et al., 2014). All these developments have brought to mind the question whether *Brucella* agents started to appear in the woods as well as the oceans (Pappas, 2010).

The studies of DNA-DNA hybridization have illustrated that there is more than 90% homology between recognized *Brucella* species (Cloeckaert et al., 2002). For this reason, in 1985, combining the 6 species into one single species, which is *B. melitensis*, and recognizing the other species as biovars, was suggested (for example, *B. melitensis* biovar abortus 1) (Cloeckaert et al., 2002; Godfroid et al., 2011; Pappas, 2010; Verger et al., 1985). However, in 2003, the subcommittee on *Brucella* taxonomy decided to go back to the previous version of the taxonomic treatment of *Brucella* genus and approved once again the six classical *Brucella* nomenspecies with their corresponding biovars (Pappas, 2010).

It is known that *Brucella* spp. can persist in milk for a period of time (even when the milk turns sour) and they continue growing in soft fresh cheese. They may also persist in ice-cream for weeks and in butter for months. Thus, it is suggested that these products should be made from pasteurized milk. What is more, *Brucella* spp. can also persist in soil and tap water (Godfroid et al., 2005). In moist soil and in dung, *Brucella* spp. can persist for 70 - 80 days. It is also stated that *B. melitensis* can persist for 15 - 40 days in dust but it depends on the humidity (Doganay & Aygen, 2003). Another point is that hard cheese made through lactic and propionic fermentation is less risky because *Brucella* agents die fast when acidity decreases below pH 4, and they die very fast when it drops below pH 3.5. It has also been pointed out that yoghurt and sour milk are less dangerous. For this reason, soft cheese made from fresh milk can include a lot of *Brucella* organisms and that is why, the products prepared from untreated milk should be avoided (Corbel, 2006). The increase of sodium chloride (NaCl) in dairy products can inhibit *Brucella* spp. (Rowe, 2014).

BRUCELLOSIS IN ANIMALS

It is stated that ingestion of bacteria coming from diseased animals or contaminated feedstuffs lead to most infections. Respiratory exposure and contamination of abraded skin and mucosal surfaces are also the other sources of the infection (Corbel, 2006). On the other hand, *Brucella melitensis*, *B. abortus* and *B. suis* might also be transmitted between animals vertically (Aparicio, 2013). The nasal and oral mucosae, where the organisms are flown into the regional lymph nodes from are the most important points of entry for bacteria. A bacteraemia might be formed from these focal points and the organisms might be situated in other lymphoid tissues such as the spleen and the iliac, mesenteric and supramammary lymph nodes, in which they may bring about a granulomatous reaction. Another bacteraemia leads to a general infection which influences some other organs such as the pregnant uterus, the udder, and their lymph nodes. Uterine lesions are mainly situated in the placentomes. Aborted fetuses and fetal membranes include a high amount of bacteria; thus, they lead to the contamination of the environment, which creates a high risk of infection of sensitive hosts (Poester et al., 2013).

Many factors such as vaccination, age, sex and herd size and density determine whether the disease is severe or not. Abortions stemming from brucellosis are more common in animals which are not vaccinated and there are more organisms shed (Corbel, 2006). More than 85% of *Brucella* might be found in the cotyledons, placental membranes, and amniotic liquid, which might reach up to 1×10^{10} colony-forming units (CFU)/ml in allantoic fluid and 1×10^{13} CFU/g of tissue in the cotyledons in ruminants, which are pregnant (Poester et al., 2013; Yumuk & O'Callaghan, 2012)

The excretion of the organisms in uterine discharges and in milk are considered as the main characteristics of the disease (OIE 2009a; OIE, 2012). Despite the fact that infected animals may abort only once, they remain carriers till the end of their lives. In the meantime, they expel many organisms in their milk and in the following parturitions, which are apparently normal (Yumuk & O'Callaghan, 2012). All infected goats do not abort; however, they can all shed *Brucella* into the environment. Under normal conditions, the main way of infection is not venereal transmission but artificial insemination with contaminated semen might lead to infection (Poester et al., 2013). When it was found out that healthy goat can carry the disease, it was considered to be one of the most important discoveries carried out in the field of epidemiology (Seleem et al., 2010).

It was pointed out that consumption of feed which was contaminated by birth and/or abortion products and uterine discharges are the main ways of the transmission of the disease in pigs. Pigs can eat fetuses which are aborted and also placental membranes. Transmission of brucellosis during copulation is very common and this process is influential on artificial insemination. *Brucellae* may be excreted in the semen by the boar even though there is nothing abnormal in their sexual organs and it does not interfere with their sexual activity (OIE 2009b; Poester et al., 2013). Infected boar can shed 10^4 - 10^5 CFU/ml of semen, which makes venereal transmission an important way of infection (Perez-Sancho et al., 2015).

All the information regarding animal brucellosis is present on the website of the Office International des Epizooties (OIE) (http://www.oie.int) and it can be accessed via the HandiStatus tool (i.e., Help with World Animal Disease Status, version 2). The information of this website is updated regularly depending upon emergency and based on the monthly and annual reports which are sent to OIE by the veterinary administrations of the countries and other official sources (Godfroid et al., 2005). Production systems, agro-ecological zones, husbandry practices, contact with wildlife, and management factors are the main factors, which affect the prevalence of Brucellosis (Godfroid et al., 2011).

Among the pre-defined 6 classical *Brucella* species, except for *B. canis*, *B. ovis* and *B. neotomae*, all these species of *Brucella* form polycyclic chain of infection and they might be transmitted to other animal species and to man when the necessary ecological-epizootic conditions are suitable (Thimm, 1982). In spite of the successful results obtained in most of the industrialized countries, in which bovine brucellosis is no longer a big problem, small ruminant brucellosis is still an important issue

(Godfroid et al., 2005). It is widely known that in the Mediterranean region, *B. melitensis* is endemic but it is a widespread worldwide disease. North America including Australia, New Zealand, South-East Asia, Northern and Central Europe (except for Mexico) are thought to be free from the agent (OIE, 2009a). In some Mediterranean countries such as Israel, Kuwait, Saudi Arabia, and Egypt, *B. melitensis* infection in cattle is particularly important and they exemplify *Brucella* agents' potential of causing infection in other hosts as well as its main hosts (Corbel, 1997; Perez-Sancho et al., 2015).

In addition to cattle and small ruminants, it was also reported that brucellosis was identified in both one-humped (Camelus dromedarius) and two-humped (C. bactrianus) camels as well as South American camelids, llama (*Lama glama*), alpaca (*L. pacos*), guanaco (*L. guinicoe*), and vicuna (*Vicugne vicugne*) (OIE, 2012; WHO, 1986). According to Saleem et al. (2010), when camels are pastured together with infected sheep and goats, they can be infected by *B. abortus* and *B. melitensis.* Although it has been belittled in the Middle East, milk coming from infected camels is one of the most important sources of infection. In some countries, camel brucellosis stemming from *B. melitensis* and *B. abortus* infections is a major source of human infections (Godfroid et al., 2005).

Infected ewes might shed *B. ovis* through their vaginal discharges and their milk. In accordance with this, the transmission from ewe to lamb and the transmission from ewe to ram might be considered as the main source of infection. For this reason, it is important for researchers to consider ewes in accordance with the infection's epidemiology, which is necessary for the effective eradication of *B. ovis* infection. It has also been reported that infected rams might shed *B. ovis* in their semen for years (OIE, 2015a). The domestic goat, moufl on (*Ovis musimon*), white-tailed deer (*Odocoileus virginianus*) and red deer (*Cervus elaphus*) are also open to infection despite the fact that the most important host species for *B. ovis* is sheep. As stated by (Poester et al., 2013), *B. ovis* is very frequently observed in coutries in which sheep raising is common, and which have flock prevelance ranging from 9.1% to 46.7%. *B. ovis* has been observed in the countries of America and Europe and also, in New Zealand and South

Africa and it is probable that this disease also occurs in other countries where sheep-raising is common (OIE, 2015a). The prevalence of Brucellosis in pigs is generally low except for such countries as South America and South-East Asia (OIE, 2009b).

B. melitensis bv-2 has recently been reported in catfish in Egypt, which revealed that fish might also be a new source of infection (Yumuk & O'Callaghan, 2012). This fish, namely, *Clarias gariepinus*, is found in the freshwater of the Nile. It has been predicted that this fish has been infected through the meat waste of infected animals which was illegally thrown into the water. Since this fish is edible by humans, it might be a new source of entry of the pathogen to the human food chain. It has been stated that *Brucella* has started to appear in rivers after its existence in oceans and the jungle (Pappas, 2010).

Each *Brucella* type has its peculiar epidemiologic features and as new types appear, the interaction between *Brucella* genus with animals and humans has become complicated. As stated by Corbel (1997) and Godfroid et al., (2005), new *Brucella* strains and types adapt to the social and agricultural practices which are continuously changing. In order to improve the quality of studies, it is necessary to train researchers in terms of formal epidemiology and statistics. If a more integrated approach to disease surveillance, which includes human health and veterinary services is used, it could be easier to understand the dynamics of the disease within the scope of animal-human relationship and resources could be used more cost-effectively (Dean et al., 2012).

HUMAN BRUCELLOSIS

Major risks of brucellosis include contaminated food and the contact with the disease due to people's occupations. In addition, consumption of unpasteurized milk coming from cows, small ruminants or camels and also some other milk products such as cheese, sour milk are the main ways of infection. Similarly, if the meat of animals infected with brucellosis is consumed without being sufficiently cooked, it might be another source of infection (Godfroid et al., 2005; Godfroid et al., 2013; Yumuk & O'Callaghan, 2012). However, meat products are rarely the major sources of infection as they are generally cooked and in muscle tissue, the amount of bacteria is low (Doganay & Aygen, 2003; Matthew & Leland 2004). Infection may be due to the bacteria from infected animals that enter into the body via skin lesions or mucous membranes or when contaminated dust or aeresols are inhaled (Doganay & Aygen, 2003; Godfroid et al., 2005).

Even though *Brucella* is thought to be very infectious when it causes infection through the respiratory route, inhalation is not one of the primary means of infection. However, some occupations such as vets, people working in laboratories or abattoirs are the main risk groups. *Brucella* spp. are believed to be the most common pathogens acquired in laboratories and they are responsible for up to 2% of all infections coming from laboratories (Seleem et al., 2010). It is stated that infection might occur during the handling of animals, which are infected particularly when these animals are around the time of parturition (Yumuk & O'Callaghan, 2012). The occupation of vets is risky both through the infected animals and their products and through the needlestick injuries when they are inoculating animals with live *Brucella* vaccines (Madkour, 2001; Matthew & Leland, 2004).

It is important to consider evaluating the fever in travelers coming from abroad, particularly the ones who are coming from endemic areas and the ones that report that they were exposed to unpasteurized milk products due to their occupation or diets (Matthew &Leland, 2004; Ragan, 2002). In some countries, the prevalence of human brucellosis coming from dairy products is seasonal and it reaches the highest numbers usually after kidding and lambing (Seleem et al., 2010). *Brucella* in humans is related to many different factors such as geographical factors, husbandry practices, techniques of food preparation, etc. It is also stated that infection can affect all human beings regardless of their age and sex (Bosilkovski, 2015).

When the epidemiology of the disease is considered, humans are believed to be an 'epidemological cul de sac' of the infection (Thimm, 1982). The transmission of the disease from human to human is rare;

therefore, it is not significant in the epidemiology of the disease (Perez-Sancho et al., 2015). Brucellosis in human is not regarded as a contagious one (Godfroid et al., 2005). It is stated that there are very few reported cases occuring through blood transfusion, bone marrow transplant, transplacental way, breastfeeding, sexual intercourse, congenital way and laboratory infection (Godfroid et al., 2005; Godfroid et al., 2013; Poulou et al., 2006; Wang et al., 2015).

In many different geographical regions, which are traditionally regarded as endemic such as France, Israel, and Latin America, the disease is under control. However, human brucellosis now has new foci, especially in central Asia and in certain countries of the near east such as Syria, it is getting worse (Pappas et al., 2006b). In a study published in the year 2006, which outlined the global epidemiology of the disease, it is seen that Central Asia and the Middle East are the main regions of the disease. For this reason, the global map of the disease is the same in 2010 despite the new hot spots added, it even exists in regions in which the infection is least expected (Pappas, 2010). It has been pointed out that in Western Europe, where the existing surveillance systems are sufficient, it is not possible for human brucellosis to re-emerge. There are also some eradication campaigns in southern European member countries which will make the spread of the disease inside the EU impossible. (Pappas et al., 2006b). Brucellosis, which is especially caused by *B. melitensis*, is still the most common zoonotic disease worldwide (Pappas et al., 2006a). In addition, when 76 diseases and syndromes observed in animals are reviewed, brucellosis is among the top 10 of the list, which affect poor people (Dean et al., 2012).

The real incidence of human brucellosis worldwide is not known (Corbel, 1997; Doganay & Aygen, 2003). According to the numbers provided by World Health Organization (WHO), there are 500,000 reported Brucellosis cases annually worldwide (Doganay & Aygen, 2003; Perez-Sancho et al., 2015). However, it is estimated that unreported cases, which are happening because of unspecific clinical symptoms, are thought to be ten times higher (Godfroid et al., 2013) or 26 times higher (Bosilkovski et al., 2009). The incidence of the disease, which is reported in endemic areas changes from < 0.01 to > 200 per 100,000 population. The reported low value of incidence in brucellosis-endemic areas which are known might be because of the lack of surveillance and reporting (Corbel, 1997). Geographic areas, which are regarded as having high level of the prevalence of brucellosis include the Middle East, the Mediterranean Basin (Portugal, Spain, Italy, Greece, Turkey, Near East, North Africa), South and Central America, South-Eastern Europe, Asia, Africa, and the Caribbean (Godfroid et al., 2005; Pappas et al., 2006b; Perez-Sancho et al., 2015).

Despite the fact that there are few reported cases of human-to-human brucellosis cases, human infection mainly stems from domestic and wild animal reservoirs (Godfroid et al., 2011; Godfroid et al., 2013; Thimm, 1982). As human-to-human transmission is rare, infected people do not constitute a threat to the ones around them. Thus, if the disease is eradicated from the animal reservoirs, it will also lower the incidence of human brucellosis (Perez-Sancho et al., 2015; Seleem et al., 2010). In order to overcome the problem of human brucellosis, it is necessary to control and eradicate animal brucellosis (Perez-Sancho et al., 2015).

REFERENCES

Aydın, N. (1997). Gram negatif küçük çomaklar-Brusella infeksiyonları. In:M. Arda, A. Minbay, N. Leleoglu, N. Aydın, M. Kahraman, O. Akay, K. S. Diker, *Özel Mikrobiyoloji Kitabı*, (pp. 110-124). Ankara, Türkiye:Medisan Yayınevi. [Gram negative small sticks- Brucellosis inflections. In: M. Arda, A. Minbay, N. Leleoglu, N. Aydın, M. Kahraman, O. Akay, K. S. Diker, *Special Microbiology Book*, (pp. 110-124). Ankara, Turkey: Medisan Publishing House].

Alton, G. G., Jones, L. M., Angus, R. D. & Verger, J. M. (1988). *Techniques for the Brucellosis Laboratory. Paris*: Institut National de la Recherche Agromique-INRA.

Alton, G. G. & Forsyth, J. (1996). Brucella- Chapter 28. In: S. Baron (Ed.), *Medical Microbiology*. 4th edition. Galveston (TX): University of Texas Medical Branch at Galveston.

Altoparlak, Ü. (2003). Brusellozun etiyolojisi. *Ankem Dergisi*, *17*(3) 330-332. [The ethiology of brucellosis. *Ankem Journal*, *17*(3) 330-332].

Aparicio, E. D. (2013). Epidemiology of brucellosis in domesti animals caused by Brucella melitensis, Brucella suis and Brucella abortus. *Rev Sci Tech OIE*, *32*(1), 53-60.

Bardenstein, S., Strata, V. & Banai, M. (2010). A fastidious bacteria but a virulent pathogen. *Vet J*, *184*, 126-127.

Bosilkovski, M., Dimzova, M. & Grozdanovski, K. (2009). Natural history of brucellosis in an endemic region in different time periods. *Acta Clinica Croatica*, *48*, 41-46.

Bosilkovski, M. (2015). *Microbiology, epidemiology and pathogenesis of Brucella.* Retrieved 11.23.2015, http://www.uptodate.com/contents/ microbiology-epidemiology-and-pathogenesis-of-brucella?source= related_link.

Cloeckaert, A., Vizcaino, N., Paquet, J. Y., Bowden, R. & Elzer, P. H. (2002). Major outer membrane proteins of *Brucella spp.*: past, present and future. *Vet Microbiol.*, *90*, 229-247.

Corbel, M. J. (2006). Brucellosis in Human and Animals. Geneva, Switzerland: *WHO Press.* Retrieved 24.09.2016, http://www.who.int/ csr/resources/publications/Brucellosis.pdf.

Corbel, M. J. & Banai, M. (2005). Genus I. Brucella. In: D.J. Brenner, N.R. Krieg, J.T. Staley, *Bergey's Manuel of Systematic of Bacteriology*, 2e, vol. 2, part C, 370-386. USA: Springer.

Corbel, M. J. & Brinley Morgan, M. J. (1982). Classification of the genus Brucella: the current position. *Rev Sci Tech (OIE)*, *1* (1), 281-289.

Corbel, M. J. (1997). Brucellosis: an overview. *Emerg Infect Dis.*, *3*(2), 213-221.

Cameron, H. S. & Meyer, M. E. (1958). Metabolic Studies on Brucella Neotomae (Stoenner and Lackman). *J Bacteriol.*, *76*(5), 546-548.

Cevik, A. M. (2001). Bruselloz Epidemiyolojisi. *ANKEM Dergisi*, *15*(3), 568-570. [Epidemiology of Brucellosis. *ANKEM Journal, 15*(3), 568-570].

Celebi, S. (2003). Bruselloz Epdemiyolojisi. *ANKEM Dergisi*, *17*(3), 340-343. [Epidemiology of Brucellosis. *ANKEM Journal, 17*(3), 340- 343].

D'anastasio, R., Staniscia, T., Milia, M. L. & Capasso, L. (2011). Origin, evolution and paleoepidemiology of brucellosis. *Epidemiol Infect*, *139*, 149–156.

Dean, A. S., Crump, L., Greter, H., Schelling, E. & Zinsstag, J. (2012). Global Burden of Human Brucellosis: A Systematic Review of Disease Frequency. *PLOS Neglected Tropical Diseases*, *6*(10), 1-9.

DelVecchio, V. G. & Mujer, C. V. (2013). Brucella spp. In: *Encyclopedia of Medical Genomics and Proteomics*. New York: Taylor and Francis.

Doganay, M. & Aygen, B. (2003). Human brucellosis: an overview. *International Journal of Infectious Diseases*, *7*, 173-182.

Gorvela, J. & Moreno, E. (2002). Brucella intracellular life: from invasion to intracelullar replication. *Vet Microbiol.*, *90*, 281–297.

Godfroid, J., Cloeckaert, A., Liautard, J. P., Kohler, S., Fretin, D., Walravens, K., Garin-Bastuji, B. & Letesson, J. J. (2005). From the discovery of the Malta fever's agent to the discovery of a marine mammal reservoir, brusellosis has continuously been a re-emerging zonosis. *Vet Res*, *36*, 313–326.

Godfroid, J., Scholz, H. C., Barbier, T., Nicolas, C., Wattiau, P., Fretin, D., et al. (2011). Brucellosis at the animal/ecosystem/human interface at the beginning of the 21st century. *Prev Vet Med.*, *102*, 118-131.

Godfroid, J., Sascha, A. D., Pappas, G., Rothf, F., Matopeg, G., Mumah, J., et al. (2013). A "One Health" surveillance and control of brucellosis in developing countries: Moving away from improvisation. *Comp Immunol Microbiol Infect Dis.*, *36*, 241– 248.

Golem, S. B. (1949). Brusellozun Memleketimizdeki Durumu. *Türk Hijyen Tecrübi Biyoloji Dergisi*, *9*, 32-63. [The Status of Brucellosis in Our Country. *Turkish Journal of Hygiene, 9,* 32- 63].

International Committee of Sytematics of Prokaryotes (ICSP). (2014). Subcommittee on the Taxonomy of Brucella. Retrieved 11.20.2015, http://www.icsp.org/subcommittee/brucella/.

Iyisan, A. S., Akmaz, Ö., Gökçen Düzgün, S., Ersoy, Y., Eskiizmirliler, S., Güler, L. & Yurtalan, S. (2000). Türkiye'de Sığır ve Koyun Brucellosis'in Seroepidemiyolojisi. *Pendik Veteriner Mikrobiyoloji Dergisi, 31*(1) 21-75. [Brucellosis in Cattle and Sheep Seroepidemiology in Turkey. *Pendik Journal of Veterinary Microbiology, 31*(1) 21-75].

Jahans, K., Foster, G. & Broughton, E. (1997). The characterisation of Brucella strains to marine Brucella isolated marine mammals. *Veterinary Microbiology*, *57*(4), 373-382.

Karagul, M. S., Baklan, E. A., Saytekin, A. M., Altuntas, B., Yildiz-Oz, G. & Erdenlig Gurbilek, S. (2017). Biovar distribution of Brucella strains isolated from livestock in Turkey between 2010-2015. *Indian J Anim Res.*, 1-6.

Matthew, L. L. & Leland, S. R. (2004). Brucellosis. *Infect Dis Clin Pract*, *12*, 7-14.

Madkour, M. M. (2001). *Madkour's Brucellosis*. Berlin: Springer.

Nicoletti, P. (2002). A short history of brucellosis. *Vet Micriobiol.*, *90*, 5-9.

Ozcanarslan, F. C. (2011). 2004-2010 Yılları Arasında Çocuk Enfeksiyon Hastalıkları Kliniğinde İzlenen Bruselloz Tanılı Çocuk Hastaların Retrospektif Olarak Değerlendirilmesi ve Yakın Temaslı Aile Bireylerinin Brucella Enfeksiyon Gelişimi Açısından İncelenmesi. Tıpta Uzmanlık, Çukurova Üniversitesi, Çocuk Sağlığı ve Hastalıkları Anabilim Dalı. [Retrospective evalution of pediatric Brucella patients followed at pediatric infection disease clinic between years 2004-2010 and screening the households members of acute patients for occurring Brucella infection. Specialization in Medicine, Çukurova University, Department of Pediatrics].

Pappas, G. (2010). The changing Brucella ecology: novel reservoirs, new threats. *Int J Antimicrob Agents.*, *365*, 8-11.

Pappas, G., Papadimitriou, P., Akritidis, N., Christou, L. & Tsianos, E. (2006b). The new global map of human brucellosis. *Lancet Inf Dis.*, *6*(2), 91-9.

Pappas, G., Panagopoulou, P., Chistou, L. & Akritidis, N. (2006a). *Brucella* as a biological weapon. *Cell. Mol. Life. Sci.*, *63*, 2229-2236.

Parascandola, J. (1998). Alice Evans, an early woman scientist at NIH. *Public Health Rep.*, *113*(5), 472-4.

Perez-Sancho, M., Garcıa-Seco, T., Domınguez, L. & Alvarez, J. (2015). *Control of Animal Brucellosis — The Most Effective Tool To Prevent Human Brucellosis*. Retrieved, 09.24.2016, http://www.intechopen.com/:http://www.intechopen.com/books/updates-on-brucellosis/control-of-animal-brucellosis-the-most-effective-tool-to-prevent-human-brucellosis.

Poester, F. P., Samartino, L. E. & Santos, R. L. (2013). Pathogenesis and pathology of brucellosis in livestock. *Rev Sci Tech (OIE)*, *32*(1), 105-115.

Poulou, A., Markou, F., Xipolitos, I. & Skandalakis, P. (2006). A rare case of *Brucella melitensis* infection in an obstetrician during the delivery of a transplacentally infected infant. *Journal of Infection*, *53*, 39-41.

Ragan, V. E. (2002). The Animal and Plant Health Inspection Service (APHIS) brucellosis eradication program in the United States. *Vet Microbiol.*, *90*, 11-18.

Robinson, A. (2003). Guidelines for coordinated human and animal brucellosis surveillance. *FAO Animal Production and Health Paper 156.* Retrieved, 08.15.2016 http://www.fao.org/3/a-y4723e.pdf.

Rowe, M. T. (2014). Brucella- Problems with dairy products. In: C. A. Batt, *Encyclopedia of Food Microbiology*, (pp. 340-343). Second Edition. London: Academic Press.

Scholz, H. C., Nöckler, K., Göllner, C., Bahn, P., Vergnaud, G., Tomaso, H. Dahouk, S. A., Kampfer, P., Cloeckaert, A., Maquart, M., Zygmunt, M. S., Whatmore, A. M., Pfeffer, M., Huber, B., Busse, H. J. & De, B. K. (2010). Brucella inopinata sp. nov., isolated from a breast implant infection. *Int J Syst Evol Microbiol.*, *60*, 801–808.

Seleem, M. N., Boyle, S. M. & Sriranganathan, N. (2010). Brucellosis: A re-emerging zoonosis. *Vet Microbiol.*, *140*(3-4), 392-398.

Seleem, M. N., Boyle, S. M. & Sriranganathan, N. (2008). Brucella: A pathogen without classic virulence genes. *Vet Microbiol.*, *129*, 1-14.

Songer, J. G. & Post, K. W. (2012). *Brusella* cinsi. In: O. Ang, Y. Ozgur, (Eds.), *Veteriner Hekimlik Mikrobiyolojisi-Hayvan Hastalığı Etkeni Olan Bakteriler ve Mantarlar.*, (pp. 200-207). İstanbul, Türkiye: Nobel Matbacılık. [Brucella genus. In: O. Ang, Y. Ozgur, (Eds.), *Veterinary Medicine Microbiology-Bacteria and Fungi with Animal Disease.* Istanbul, Turkey: Nobel Publishing.

The Center for Food Security & Public Health (CFSPH). (2012). *Canine Brusellosis: Brucella canis.* Retrieved 11.19.2015, http://www.cfsph.iastate.edu/Factsheets/pdfs/brucellosis_canis.pdf.

Thimm, B. (1982). *Brucellosis- Distribution in Man, Domestic and Wild Animals*. Berlin: Springer.

Tiller, R. V., Gee, J. E., Lonsway, D. R., Gribble, S., Bell, S. C., Jennison, A., Bates, J., Coulter, C., Hoffmaster, A. R. & De, B. K. (2010). Identification of an unusual Brucella strain (BO2) from a lung biopsy in a 52 year-old patient with chronic destructive pneumonia. *BMC Microbiol.*, 10-23.

Verger, J., Grimont, F., Grimont, P. A. & Grayon, M. (1985). *Brucella*, a monospecific genus as shown by deoxyribonucleic-acid hybridization. *Int J Syst Bacteriol.*, *35*, 292-295.

Vicente, A. F., Antunes, J. M., Lara, G. H., Mioni, M. R., Allendorf, S. D., Peres, M. G., Appolinario, C. M., Listoni, F. J. P., Ribeiro, M. G. & Megid, J. (2014). Evaluation of three formulations of culture media for isolation of Brucella spp. regarding their ability to inhibit the growth of contaminating organisms. *Biomed Res Int.*, *702072*, 1-3.

Whatmore, A. M., Davison, N., Cloeckaert, A., Al Dahouk, S., Zygmunt, M. S., Brew, S. D., Perrett, L. L., Koylass, M. S., Vergnaud, G., Quance, C., Scholz, H. C., Dick, E. J., Hubbard, G., Schlabritz-Loutsevitch, N. E., et al. (2014). Brucella papionis sp. nov., isolated from baboons (Papio spp.). *Int J Syst Evol Microbiol.*, *64*, 4120–4128.

World Health Organisation. (2004). *Laboratory safety manual*, Third edition. Geneva.

World Health Organisation for Animal Health (OIE). *Terrestrial Manual* Chapter 2.7.2. (2009). Caprine and ovine brucellosis. Retrieved, 09.24.2016, http://web.oie.int/eng/normes/MMANUAL/2008/pdf/2.07.02_CAPRINE_OVINE_BRUC.pdf.

World Health Organisation for Animal Health (OIE). *Terrestrial Manual* Chapter 2.4.3. (2012). Bovine brucellosis. Retrieved, 09.24.2016, http://web.oie.int/eng/normes/MMANUAL/2008/pdf/2.04.03_BOVINE_BRUCELL.pdf.

World Health Organisation for Animal Health (OIE). *Terrestrial Manual* Chapter 2.1.4. (2016). Brucellosis (B. abortus, B. melitensis, B. suis) (Infection with B. abortus, B. melitensis, B. suis). Retrieved, 09.24.2016, http://www.oie.int/fileadmin/Home/eng/Health_standards/tahm/2.01.04_BRUCELLOSIS.pdf.

World Health Organisation for Animal Health (OIE). *Terrestrial Manual* Chapter 2.7.9. (2015a). Ovine Epididymitis (Brucella ovis) Retrieved, 09.24.2016, http://web.oie.int/eng/normes/MMANUAL/2008/pdf/2.07.09_OVINE_EPID.pdf.

World Health Organisation for Animal Health (OIE). *Terrestrial Manual* Chapter 2.8.5. (2009b). Porcine Brucellosis. Retrieved, 09.24.2016 http://web.oie.int/eng/normes/MMANUAL/2008/pdf/2.08.05_PORCINE_BRUC.pdf.

Wyatt, H. V. (2013). Lesson from the history of brucellosis. *Rev Sci Tech OIE*, *32*(1), 17-25.

Young, D. A. (1995). Florance Nightingale's fewer. *BMJ*, *311*, 1697-1700.

Yumuk, Z. & O'Callaghan, D. (2012). Brucellosis in Turkey-an overview. *Int J Infect Dis.*, *16*(4), 228-35.

In: Advances in Health and Natural Sciences ISBN: 978-1-53614-639-4
Editors: Burcu Yuksel et al.

Chapter 9

MUTAGENIC POTENTIAL OF THE INSECTICIDE ARRIVO 25 EC ON MITOSIS AND MEIOSIS OF *HYACINTHUS ORIENTALIS*

Ozlem Aksoy[1,*], Aylın Horozal[1] and Burcu Yuksel[2]
[1]Department of Biology, Faculty of Arts and Sciences,
Kocaeli University, Kocaeli, Turkey
[2]Kocaeli Vocational School of Health Services,
Kocaeli University, Kocaeli, Turkey

ABSTRACT

The effects of Arrivo 25 EC (250g/l, Cypermethrin), a pyrethroid insecticide, were investigated on *Hyacinthus orientalis* grown in greenhouse. The insecticide was applied to *H.orientalis* at recommended dosage and double the recommended dosage by manufacturing company. Mitosis was examined in root tip cells and meiosis was examined in anthers. It was observed that the insecticide used in trial applications caused abnormalities in mitosis and pollen meiosis in parallel with the

* Corresponding Author Email: odalgic@yahoo.com.

increase in dosage. It is obvious that this condition will have a negative effect on pollen fertility.

Keywords: cypermethrin, *Hyacinthus orientalis,* meiosis, mitosis, mutagenic

INTRODUCTION

Pesticide applications in agricultural and ornamental plants have harmful effects on their cytological mechanisms. Cypermethrin is a synthetic pyrethroid insecticide used to control many pests, including moth pests of cotton, fruit and vegetable crops (Meister, 1992). It is also used for crack, crevice and spot treatment for control of insect pests in stores, warehouses, industrial buildings, houses, apartment buildings, greenhouses, laboratories and on ships, railcars, buses, trucks and aircraft. It may also be used in non-food areas in schools, nursing homes, hospitals, restaurants, hotels, and in food processing plants and as a barrier treatment insect repellent for horses (USEPA, 1989). Cypermethrin is a moderately toxic material by dermal absorption or ingestion (Meister, 1992; OHS, 1993). It may cause irritation to skin and eyes. Symptoms of dermal exposure include numbness, tingling, itching, burning sensation, loss of bladder control, incoordination, seizures and possible death. Pyrethroids may adversely affect the central nervous system (OHS, 1993). Cypermethrin is extremely toxic to fish and aquatic invertebrates. The 96-hour LC50 for cypermethrin in rainbow trout is 0.82 ppb, and in bluegill sunfish is 1.78 ppb. Its acute LC50 for Daphnia magna, a small freshwater crustacean, is 0.2 ppb. The bioconcentration factor for cypermethrin in rainbow trout was 1200 times in a flow through study (USEPA, 1989). If exposed to cypermethrin during pregnancy, rats give birth to offspring with developmental delays. In male rats exposed to cypermethrin, the proportion of abnormal sperm increases. It also causes genetic damage: chromosomal abnormalities increased in bone marrow and spleen cells when mice were exposed to cypermethrin (Amer et al., 1993). One study showed that cypermethrin inhibited gap junctional intercellular communication, which plays an important role in cell growth and is inhibited by carcinogenic agents (Tateno et al., 1993). Studies have shown that residue from cypermethrin can last for 84 days in the air, on walls, floor and furniture (Wright et al., 1993). Cypermethrin is an environment pollutant and, due to its use, toxicity and persistence, it may cause significant damage to non-target organisms and various ecosystems. This necessitates the development of an effective tool to eliminate or obstruct the concentration of this pesticide (Gangola et al., 2018). Insecticides affect the plants by causing harm to their metabolism, photosynthesis, development and yield. To give an example, malathion formation and carbarly formulations obstruct the photosynthesis of field grown soybean or the four formulation insecticides, namely flucythrinate, chlordimeform, carbaryl and malathion, lessen the stomatal conductance and lint yield.

The effects of insecticides on growth, fruiting patterns, and yield of Acala-SJ-2 cotton were observed in Wen, et al. (2017). In another study, Sheo Mohan Prasad et al. (2015) found that the decrease in photosynthetic rate caused decreased growth of *A. pinnata* at higher concentrations of both chlorpyrifos and cypermethrin.

Pollen germination was reduced by *in vitro* fungicide applications and degeneration in pollen tubes occurred (Pavlik & Jandurova 2000). Tort et al. (2005) observed that pyrimethanil fungicide caused changes in the structure of tomato pollens. Moreover, it was reported that fungicides resulted various chromosomal abnormalities, such as Deltan applications (0.3%) gave rise to chromatid and subchromatid breaking in red pepper (Prakash & Prathapasenan, 1988). Thiodan, Folithion, Lebaycid and Kitazin caused fragmentation, lagging chromosomes, bridges, tripolar and tetrapolar spindle in barley (Grover & Tyagi, 1980). Therefore, it is necessary to study the effect of overdose application of common pesticides on plants. In this study, the effects of different concentrations recommended and double of the recommend dosages of Arrivo 25 EC on mitosis and meiosis of *H.orientalis* were investigated.

MATERIALS AND METHODS

Plant Material and Arrivo 25 EC

The selected plant material healthy *H.orientalis* bulbs were obtained from the market. The insecticide used in the study was Arrivo 25 EC (250g/l Cypermethrin, $C_{22}H_{19}Cl_2NO_3$), which was produced and distributed by the company Hektaş.

Application of Arrivo 25 EC

The commercially maintained and selected *H.orientalis* bulbs were germinated for primary roots of 1–2 cm long in erlenmeyer flasks containing distilled water. The germinated primary roots were treated with three different concentrations of (%0.5, %1, %2.5 ml) Arrivo 25 EC. All experiments were carried out in three repetitions.

Mitotic Analysis

The flower buds were fixed first in Carnoy solution (3:1 alcohol-acetic acid) and hydrolyzed in 1 N HCl at 60°C for 5–10 minutes. They were, then, crushed in a 2% orcein stain in 45% acetic acid and the slides were examined immediately. The mitotic frequency was determined by examination of 1000 cells per slide.

Detection of Chromosomal Abnormalities

To investigate the abnormality percentages (chromosomal bridges, lagging chromosomes, chromosomal fragments, etc.) in different stages of mitosis and meiosis I, II, abnormal cells were counted and the percentage of abnormal cells was determined in each phase. Chromosomal abnormalities were evaluated separately in all phases of mitosis and photographic data was obtained using an Olympus BX51 photomicroscope.

Statistical Analysis

The statistical analyses of data were carried out using SPSS for Windows version 11.0 statistical software (SPSS Inc., Chicago, USA). Statistically significant differences between the groups were compared using both one-way analysis of variance (ANOVA) and Chi-Square Test. The data were displayed by mean ± standard deviation (SD). P-values less than 0.05 were considered statistically significant.

Results and Discussion

Mitotic index (MI), which reflects the frequency of cell division, is an acceptable measure of cytotoxicity for all living organisms. If the mitotic index decreases below 22% of the control, it can cause lethal effects on the test organism (Cavusoglu et al., 2011). This study demonstrated the use of higher concentrations of each different Arrivo 25 EC concentration in the various phases of mitosis. Cytotoxicity was defined as decrease in mitotic index revealed the mitodepressive effect of Arrivo on mitotic index (MI) of *H.orientalis*. While treating seeds with different concentrations of Arrivo 25 EC, we observed that Mitotic Frequency Percentage decreased while the concentrations increased for each concentration of Arrivo 25 EC. The mitotic index of 0.5 ml and 2.5 ml % concentrationswas slightly higher compared to control. The Arrivo 25 EC concentrations used had a negative effect on mitotic division. The mitotic frequency was 15.3% in the control group and decreased to 12.9% at 0.1ml concentration, 10.8% 0.5 ml concentration and 8.0% at 2.5 ml concentration Arrivo 25 EC (Table 1). The Arrivo 25 EC concentrations used had a negative effect on mitotic division. In high Arrivo 25 EC concentration, mitotic frequency reduced to almost half of that of the control (Figure 1), which affirmed the relationship between this parameter and morphological characters as previously reported by Abderrhman (1998), Abou Elkheir (1992) and Alam et al. (1987). This increase may due to shortening the duration of mitotic cycle and enhancing the interphase cells that entered the subsequent division stages (Haroun & Al shehri 2001) or because of the prompting of the synthesis of DNA in dividing cells (Badr et al., 1985; Chand & Roy 1981). Inhibition of mitotic activities was often used for tracing cytotoxic substances.

Table 1. The effect of Arrivo 25 EC concentrations on mitotic frequency percentage

Concentration (ml)	Cell numbers	Prophase	Metaphase	Anaphase	Telophase	Divided Number of Cells	Mitotic Frequency Percentage (%)
Control	1000	79	16	21	37	153	15.3
0.1	1000	58	21	13	37	129	12.9
0.5	1000	53	15	13	27	108	10.8
2.5	1000	45	11	11	13	80	8.0

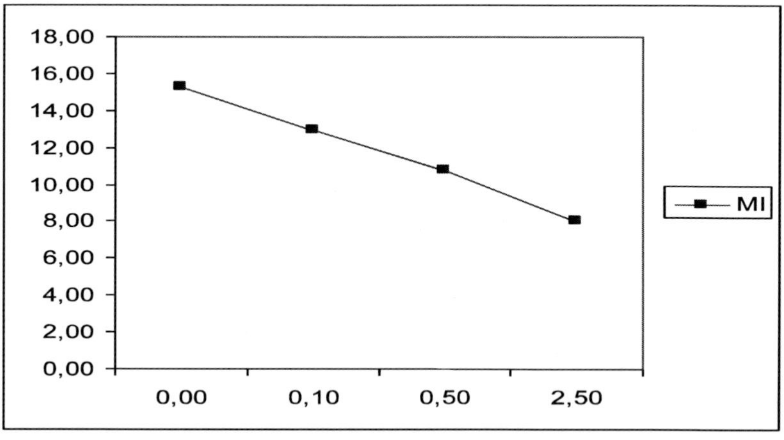

Figure 1. Mitotic frequency percentage in *H.orientalis* treated and control groups in root meristematic cells.

We also examined the mitotic abnormality percentages and frequencies for prophase, metaphase, anaphase and telophase for different concentrations at different hours (Table 2). We observed that mitotic abnormality increased for all measurements as the concentration amount increased. For example, for prophase, the abnormality percentage was 0% in the control but increased to 66.6% at 0.1 ml, 66.6% at 0.5 ml and 2.5 and 100% at 2.5 ml concentration Arrivo 25 EC.

Table 2. The effect of Arrivo 25 EC concentrations on the mitotic abnormality percentage and frequencies

Concentration	Prophase I			Metaphase I			Anaphase I			Telophase I		
%ml	N	A	%A	N	A	%A	N	A	%A	N	A	%A
Control	2	-	-	3	79	96.3	-	4	100	2	-	-
0.1	1	2	66.6	3	33	91.6	-	3	100	2	3	60
0.5	2	4	66.6	3	4	57.1	1	8	88.8	2	-	-
2.5	-	2	100	-	8	100	-	4	100	-	1	100

N; Normal cell, A; Abnormal cell.

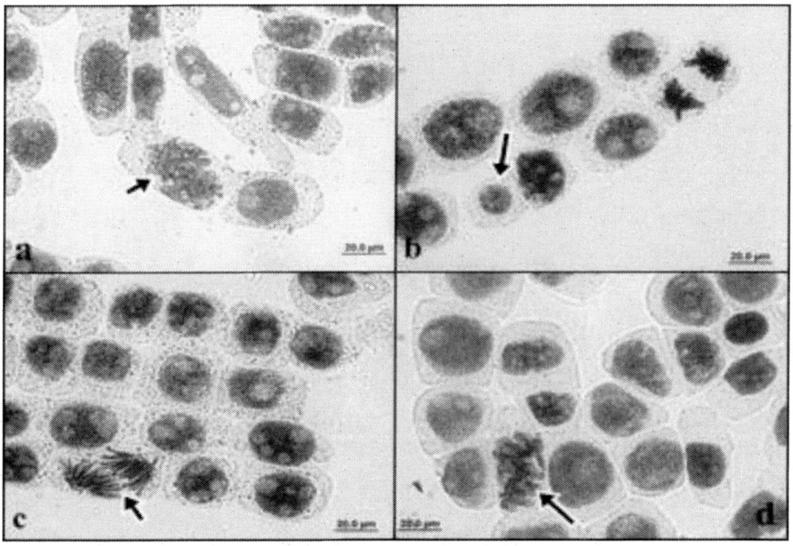

Figure 2. The following common abnormalities were observed at 0.5 ml Arrivo 25 EC; a. Chromosomal bridge in telophase, b. Picnotic nucleus, c. Multipolar anaphase d. Multipolar metaphase. The abnormalities present are indicated by arrows.

It was determined that the chemical treatment (including Arrivo) had drastic impact on non-target insect species (Lovei & Sunderland, 1996). Detrimental effects of insecticides, acaricides or other chemical compounds on pollen germination, fruit set and/or yield were reported on apple (Mayer & Lunden, 1986; Bound & Jones, 2004; Nichols et al., 2004). Mayer and Lunden (1986) noted that acaricides used during bloom influenced pollen viability of apple and pear.

The 0.5 ml Arrivo 25 EC concentration were found in *H. orientalis* Chromosomal bridge in telophase, Picnotic nucleus, Multipolar anaphase, Multipolar metaphase (Figure 2).

The 2.5 ml Arrivo 25 EC concentration were found in *H. orientalis* Chromosomal bridge in telophase, Laggard chromosome in anaphase, Normal mitosis, Multipolar anaphase (Figure 3).

The 2.5 ml Arrivo 25 EC concentration were found in *H. Orientalis* at meiosis I Laggard chromosomes in interphase, abnormal prophase, chromosomal bridge in anaphase, chromosomal bridge and micronucleus formation in telophase (Figure 4).

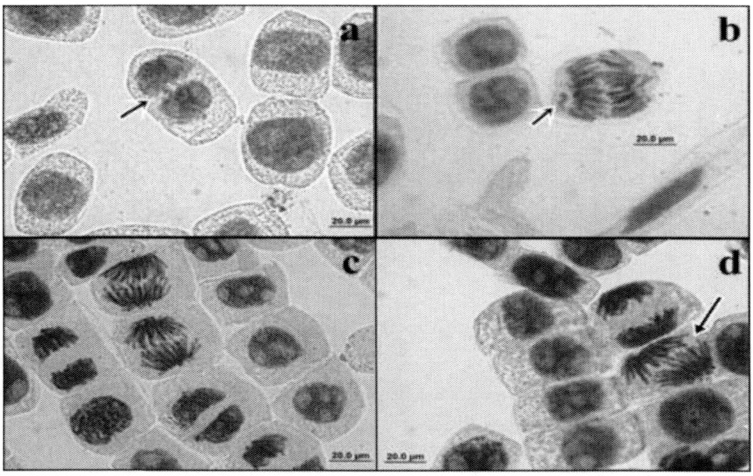

Figure 3. The following common abnormalities were observed at 2.5 ml Arrivo 25 EC; a. Chromosomal bridge in telophase, b. Laggard chromosome in anaphase, c. Normal mitosis, d. Multipolar anaphase. The abnormalities present are indicated by arrows.

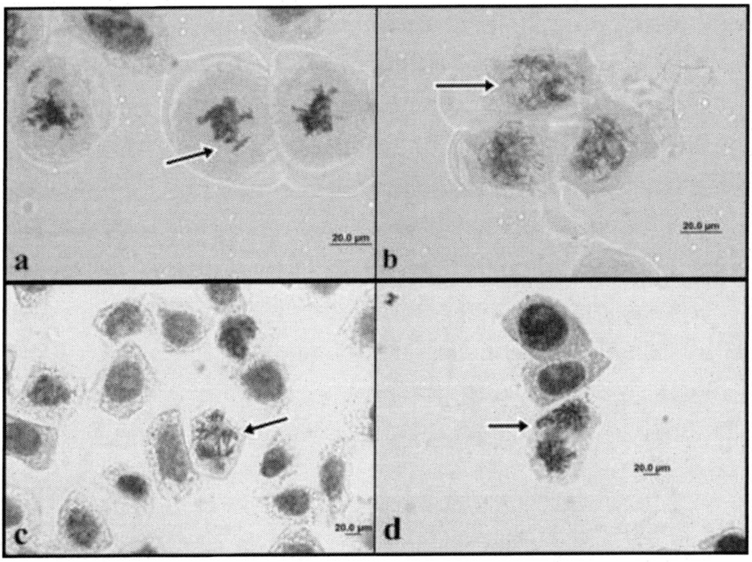

Figure 4. The following common abnormalities were observed at 2.5 ml Arrivo 25 EC at meiosis I. A. Laggard chromosomes in interphase, b. Abnormal prophase, c. Chromosomal bridge in anaphase, d. Chromosomal bridge and micronucleus formation in telophase. The abnormalities present are indicated by arrows.

The 2.5 ml Arrivo 25 EC concentration were found in *H. Orientalis* at meiosis II Picnotic nucleus, deformations in nuclei, chromosomal bridge in anaphase, micronucleus, Laggard chromosome, chromosome decondensation and micronuclei formation (Figure 5).

To investigate the effects of toxic substances applied to plant organisms, it is important to investigate the chromosome anomalies and mitotic index in the root meristematic cells. The mitotic index reflects the frequency of cell division and is used as an important parameter in determining the rate of root development (Jiang et al., 2000). The mitotic index decrease is defined as the lethal effect (Antonsie-wiez, 1990), which falls below 22% when compared with control groups, and the sublethal effect, which falls below 50% (Panda & Sahu, 1985). These values were defined by the researchers as cytotoxic limit values (Sharma, 1983). Schulze and Kirscher (1996) have claimed that the mitotic index depression is due to the fact that genotoxins with mitochondrial deprivation have the power to block DNA and nucleoprotein synthesis. El-Ghamery et al. (2000) demonstrated that mitochondrial inhibition is caused by blockage of G1 phase and suppression of DNA synthesis in S phase. Based on the results of their study, Bal (1995) suggested that mitotic index depression may be due to mitotic cycling, as a physiological response to unprotected cells against toxic chemical agents such as energy pathway, inhibition of protein, RNA and DNA synthesis. Kaymak (2005) suggested that various chemicals might be the result of the influence of microtubules on the chromosomes, which are essential proteins that provide the formation of spindle fibers necessary for mitosis. Eun et al. (2000) noted that chemical agents altered microtubule organization and indirectly inhibited mitotic activity.

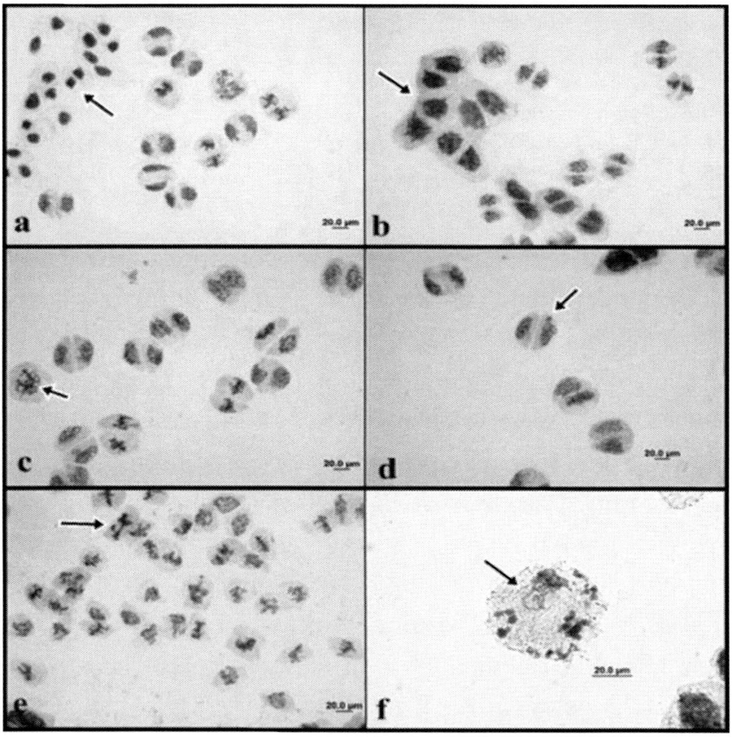

Figure 5. The following common abnormalities were observed at 2.5 ml Arrivo 25 EC at meiosis II. a. Picnotic nucleus, b. Deformations in nuclei, c. Chromosomal bridge in anaphase, d. Micronucleus, e. Laggard chromosome, f. Chromosome decondensation and micronuclei formation. The abnormalities present are indicated by arrows.

Changes in the number and structure of chromosomes occur as a result of abnormal placement of chromosomes by the effect of toxic chemical agents. Examples of loss of chromosome include backward chromosome, adhesiveness, multipolarity and c-mitosis. Materials that cause fractures in chromosomes can cause changes in chromosome bridge formation or chromosome structure (Leme & Marin-Morales, 2009; Radic et al., 2010).

Backward chromosomes might form as a result of the disorders which occur in the organization or functions of the spindle threads (Yıldız et al., 2009; Fındıklı & Türkoğlu, 2010; Türkoğlu, 2012); anaphase bridge might form as a result of the translocation with unequal chromatin changes and presence of chromatin breakage (Liman, 2013); c-mitosis formation is seen as a result of degradation of spindle yarns (Yoshida et al., 1983); hyperhomologous chromosomes arise as a result of backwardness in chromosomes, inadequacy of chromosomal motions or irregularities in the signaling process, depending on the state of the spindle fibers being affected; multipolarity emerges as a result of partial influences of spindle fibers (Ahmad 1992); whereas adhesion in chromosomes occurs as a result of physical adhesion of chromatin proteins. Moreover, it has been reported that bridge formation may occur during chromosomal adhesions, disentric chromosome formation, merger in sister chromatids and chromosomes moving towards the poles. In addition, fragments of broken chromosomes can cause micronucleus formation and loss of genetic material (Topaktas & Rencuzogullari, 1995).

As stated before, decreases in the mitotic index, increases in the numbers of abnormally dividing cells and increases in chromosomal abnormalities are caused by the inhibition of energy pathway, protein, RNA and DNA synthesis, the alteration of the microtubule organization and the synthesis mechanism of the main proteins that provide the formation of spindle spinning, affecting genes or gene products that regulate cell division control points (Yuksel & Aksoy 2017; Aksoy et al., 2012). Based on these results, we can infer that Cypermethrin acts like a cytotoxic agent.

Occurrence of chromosomal bridges, laggard chromosomes and multipolar anaphases clearly shows the accumulated effects of Arrivo on the spindle formation. The phenomenon of chromosome stickiness indicates the toxicity of Arrivo and occurrence of breaks and bridges show that it is a potent clastogen. In sum, the present study shows that Arrivo, commercial formula of Cypermethrin, is genotoxic because this pesticide induced chromosome abnormalities in *H.orientalis*. Since, mitotic index is decreased, Arrivo has high cytotoxical effects. As a result of cytological studies, it can be said that Cypermethrin is effective both in chromosome organization and DNA level. Hence, more care should be given to the application of Arrivo in agriculture, in countries where it is widely used.

REFERENCES

Abderrhman, S. M. (1998). Cytogenetic effects of *Peganum harmala* on Maize root tips. *Cytologia*, 63: 283-291.

Abou elkheir, Z. A. & Abou elkheir, G. M. (1992). Cytological effect of certain active constituents of *Peganum harmala*. 1. Effects of hormonal and harmine alkaloids on mitosis of *Allium cepa*. *J. King Saud Univ. Science*, 4. 37-45.

Ahmad, S., & Yasmin, R. (1992). Effects of methyl parathion and tri-miltox on the mitosis of *Allium cepa*. *Cytologia*, 57: 155-160.

Aksoy, O., Erbulucu, T., Oner, S., & Tekeli, I.B. (2012). Phytotoxic and genotoxic effects of water samples taken from the eastern channel of Kocaeli on *Vicia faba* and *Zea mays*. *Fresen. Environ. Bull.*, 21(7), 1819-1826.

Alam, S. Golam, K., Amin, M., Islam, M. (1987). Mitotic effect of leaf extract of *Ipomoea carnea* on *Allium cepa*. *Cytologia*, 52: 721-724.

Amer, S. M., et al. (1993). Induction of chromosomal aberrations and sister chromatid exchange in vivo and in vitro by the insecticide cypermethrin. *Journal of Applied Toxicology* 13:341-345.

Antosiewicz, D. (1990). Analysis of the cell cycle in the root meristem of Allium cepa under the influence of ledakrin. *Folia Histochemica et Cytobiologica*, *28*(1-2), 79-95.

Badr, A. Hamoud, M. A. and Haroun, S. A. (1985). Effects of herbicide Gespax on mitosis, mitotic chromosomes and nucleic acid in *Vicia faba* L. root meristems. *Proc. Saudi boil., Soc.,* 8: 359-369.

Bound, S. A. & Jones, K. M. (2004). Ammonium thiosulphate as a blossom thinner of 'Delicious' apple, 'Winter Cole' pear and 'Hunter' apricot. *Austr. J. Exp. Agric.,* 44: 931–937.

Cavuşoğlu K, Kaya A, Yılmaz F & Yalçın E. (2011). Effects of Cypermethrin on *Allium cepa. Environ. Toxicol,* 42: 181–189.

Chand, S. & Roy, S. C. (1981). Effects of herbicide 2, 4 dinitrophenol on mitosis, DNA, RNA and protein synthesis in *Nigella sativa* L. *Biol. Plant*., 23: 198-202.

El-Ghamery, A. A., El-Nahas, A. I., Mansour, M. M. (2000). The action of atrazine herbicide as an inhibitor of cell division on chromosomes and nucleic acids content in root meristems of *Allium cepa and Vicia faba. Journal Cytologia*, 65 (3): 277-287.

Eun, S. O., Young, H. S., Lee, Y., (2000) Lead disturbs microtubule organization in the root meristem of *Zeal mays*. *Physiology Plant*., 110: 356-365.

Fındıklı Z., & Türkoğlu Ş. (2010). Glyphos ve DDVP' nin *Allium cepa* L.' da mitoz bölünme ve kromozomlar üzerine etkisi, *C.Ü. Fen-Edebiyat Fakültesi Fen Bilimleri Dergisi*, 31, 49-62. [The Effects of Glyphos and DDVP on Mitotic Division and Chromosomes in Allium cepa L., *CU Faculty of Sciences and Letters, Journal of Science, 31*, 49-62.

Grover, I. S., & Tyagi, P. S. (1980). Chromosomal aberrations induced by pesticides in meiotic cells of barley. *Caryologia*, 33(2), 251-259.

Haroun, S. A. & Al shehri A. M. (2001). Cytogenetic effects of *Calotropis procera* extract on *Vicia faba*. *Cytologia*, 66: 337-378.

Jiang, W., & Liu, D. (2000). Effects of Pb 2+ on Root Growth, Cell Division, and Nucleolus of Zea mays L. *Bulletin of environmental contamination and toxicology*, *65*(6), 786-793.

Kaymak, F. (2005). Cytogenic effects of maleic hydrazide on *Helianthus annus* L. Pakistan *Journal of Biological Sciences*., 8: 104-108.

Leme D. M., Marin-Morales M. A., (2009). Allium cepa test in environmental monitoring: A review on its application, *Mutat. Res.-Rev. Mutat.*, 682, 71-81.

Liman R. (2013). Genotoxic effects of Bismuth (III) oxide nanoparticles by *Allium* and Comet assay, *Chemosphere.*, 93, 269-273.

Lovei, G. L. & Sunderland, K. D. (1996). Ecology and behavior of ground beetles (Col. Carabidae). *Annual Review of Ent.* 41:231-256.

Mayer, D. F. & Lunden, J. D. (1986). Toxicity of fungicides and an acaricide to honey-bees (Homoptera, Apidae) and their effects on bee foraging behavior and pollen viability on blooming apples and pears. *Environmental Entomol.,* 15: 1047–1049.

Meister, R. T. (1992). *Farm Chemicals Handbook '92.* Meister Publishing Company, Willoughby, OH.

Occupational Health Services, Inc. (1993). (Nov. 17). *MSDS for Cypermethrin.* OHS Inc., Secaucus, NJ.

Pavlık, M., & Jandurová, O. M. (2000). Fungicides cytotoxicity expressed in male gametophyte development in Brassica campestris after in vitro application of converted field doses. *Environmental and experimental botany,* 44(1), 49-58.

Panda, B. B., & Sahu, U. K. (2002).Induction of abnormal spindle function and cytokinesis inhibition in mitotic cells of *Allium cepa* by the organophosphorus insecticide fensulfothion, *Cytobios*, 42, 147-155.

Prakash, L., Dutt, M., & Prathapasenan, G. (1988). NaCl alters contents of nucleic acids, protein, polyamines and the activity of agmatine deiminase during germination and seedling growth of rice (Oryza sativa L.). *Functional Plant Biology,* 15(6), 769-776.

Prasad, S. M., Singh, A., Singh, P. (2015). Physiological, biochemical and growth responses of Azolla pinnata to chlorpyrifos and cypermethrin pesticides exposure: a comparative study. *Chemistry and Ecology*, 31(3), 285-298.

Radic S., Stipanicev D., Cvjetko P., Mikelic I. L., Rajcic M. M., Sirac S., Pevalek-Kozlina B., Pavlica P. (2010). Ecotoxicological assessment of industrial effluent using duckweed (*Lemna minor* L.) as a test organism, *Ecotoxicology*, 16, 216-22.

Sharma, C. B. S. R. (1983). Plant meristems as monitors of genetic toxicity of environmental chemicals. *Current science*, 1000-1002.

Schulze, E., & Kirschner, M. (1986). Microtubule dynamics in interphase cells. *The Journal of Cell Biology*, *102*(3), 1020-1031.

Tateno, C. et al. (1993). Effects of pyrethroid insecticides on gap junctional intercellular communications in Balb/c3T3 cells by dye-transfer assay. *Cell biology and toxicology journal 9:215-222.*

Topaktaş, M. & Rencüzoğulları, E. (1995). *Sitogenetik*, Çukurova Üniversitesi Fen Edebiyat Fakültesi Biyoloji Bölümü, Adana, Türkiye, 68-113. [*Cytogenetic*, Cukurova University, Faculty of Sciences and Letters, Department of Biology, Adana, Turkey, 68- 113].

Tort, N., Öztürk, İ., Güvensen, A. (2005). Effects of some fungicides on pollen morphology and anatomy of tomato (Lycopersicon esculentum Mill.). *Pakistan Journal of Botany*, *37*(1), 23-30.

Türkoğlu Ş. (2012). Determination of genotoxic effects of chlorfenvinphos and fenbuconazole in *Allium cepa* root cells by mitotic activity, chromosome aberration, DNA content, and comet assay, *Pestic. Biochem. Phys.*, 103, 224-230.

U.S. Environmental Protection Agency (1989). *Pesticide Fact Sheet Number 199: Cypermethrin.* US EPA, Office of Pesticide Programs, Registration Div., Washington, DC.

Wen, Y. M., Meng, L. Y., Liu, S. T., Zhu, C. M., Liu, Y. J., & Qu, A. J. (2017). Ultrastructure effects of omethoate and cypermethrin insecticides on maize seedlings. *Chemistry and Ecology*, 33(9), 883-892.

Wright, C. G., R. B. Leidy, & H. E. Dupree, Jr. (1993). "Cypermethrin in the ambient air and on surfaces of rooms treated for cockroaches." *Bulletin of Environmental Contamination and Toxicology 51:356-360.*

Yıldız, M., Ciğerci İ. H., Konuk M., Fidan A. F., Terzi H. (2009). Determination of genotoxic effects of copper sulphate and cobalt chloride in *Allium cepa* root cells by chromosome aberration and comet assays. *Chemosphere*, 75, 934–938.

Yoshida, Y., Nakamura, K., & Hiura, A. (1983). Contraction of chromosomes and depression of RNA synthesis by isopropyl N-(3-chlorophenyl) carbamate (CIPC) in *Vicia faba*. *Cytologia*, 46: 169-172.

Yuksel, B., & Aksoy, O. (2017). Cytological effects of coumarin on the mitosis of *Lens Culinaris* Medik. *Fresenius Environmental Bulletin*, *26*(11), 6400-6407.

In: Advances in Health and Natural Sciences ISBN: 978-1-53614-639-4
Editors: Burcu Yuksel et al.

Chapter 10

ALUMINUM-INDUCED TOXICITY AND PROGRAMMED CELL DEATH IN PLANTS

Filiz Vardar*[*], *Fatma Yanık, Aslıhan Çetinbaş-Genç and Gamze Kurtuluş
Marmara University, Science and Arts Faculty, Department of Biology, Göztepe Campus, Istanbul, Turkey

ABSTRACT

Wild flora and agricultural crops are faced with a variety of environmental stress factors culminating in considerable economic losses worldwide. Aluminum toxicity is one of the important abiotic stresses limiting plant growth. Aluminum (Al) is the third most abundant element in the soil, after oxygen and silicone; comprising approximately 7%. Whereas it exists primarily in the form of insoluble aluminosilicates or oxides, the complex Al becomes solubilized in acidic soils (< pH 5.0). After Al absorption, the toxicity effects root growth adversely resulting in poor uptake of water and nutrients. The root apex consisting root cap, meristem and elongation zone has a major role in Al-penetration mechanism and accumulates more Al. The common effects of Al are

[*] Corresponding Author Email: filiz.vardar@gmail.com.

thick, cracked and stubby roots, inhibited cell and root elongation, increment in cell wall rigidity, callose formation, alteration of plasma membrane and cytoskeleton, production of oxygen-free radicals, and interaction with DNA. These cellular alterations lead to oxidative stress and programmed cell death. Programmed cell death (PCD) is genetically controlled and highly regulated cell disassembly, which appears during development or under stressful conditions. During PCD, some common cellular characteristics such as cell shrinkage, vacuolization, cytochrome c release from mitochondria, specific protease activation, chromatin condensation and DNA fragmentation become apparent both in plants and animals. A few researchers indicate that Al toxicity also causes PCD. It has been known that some of the plant species have self-tolerance metabolism against Al toxicity. However, Al toxicity may be alleviated by exogenous application of organic and inorganic compounds. Even if many researchers have performed to understand the mechanism of Al toxicity and tolerance, it has not yet been fully characterized because of the complicated chemical nature of Al. Moreover, the underlying molecular, genetic and physiological features are still not clear. Although this review will describe Al toxicity and PCD principles generally, our aim is to focus especially on the mechanisms of Al-induced PCD which have been largely ignored with the goal in mind for environmental monitoring.

Keywords: acid soil, aluminum toxicity, aluminum tolerance, plant, programmed cell death, root

INTRODUCTION

Although aluminum (Al) is one of the most abundant and prevalent mineral in the earth's crust (7%), it is not assumed as an essential element for living organisms (Matsumoto, 2000; Abate et al., 2013). Al, as an abiotic stress factor, is one of the major plant growth-limiting factors at low pH (acid) soils. It has been thought that 67% of the world's potential farmable lands have acid soil below pH 5.0 and face to Al toxicity (Kochian et al. 2004; Abate et al. 2013; Ma et al. 2014). Al is a non-toxic constituent in the neutral or weakly acidic soils as insoluble aluminosilicates or oxides, but when the soil turns out to be more acid, complex Al is solubilized into phytotoxic forms. $Al(H_2O)_6^{3+}$ dissolve to

Al^{3+} ion and it is the most toxic and dominant form in low pH soils (Matsumoto, 2000; Vardar and Ünal, 2007). Whereas occurrence of acidic soil is a natural occurrence in tropical and subtropical zones such as South America, Central Africa and Southeast Asia, it is also a critical environmental problem in temperate zones. In industrial regions such as USA, Canada and Europe countries environmental pollution and acid rains raise the solubilized Al^{3+} ions in the soil (Vitorello et al., 2005). Soil acidity is also attributed to exhaustive agriculture, inappropriate farming and over leaching with basic actions. As a broad notion, acid soil fertility is regressed by the reason of toxic (Al, Cd and Mn) and deficient minerals (P, Ca, Mg and Mo). It has been stated that solubilized Al presents at the range of 10–100 μM in acid soils, and regarded as a primary constraint for crop production. Researchers also evidenced that Al adversely affect plant growth and development within a few minutes even at low micro-molar doses in acid soils (Eswaran et al., 1997; Ciamporova, 2002; Vitorello et al., 2005; Vardar and Ünal, 2007; Abate et al., 2013).

Considering the whole plant, toxic Al ions have adverse effects such as structural and anatomical deformation, chlorosis and necrosis. Besides, reduction in total leaf number, decreased photosynthetic activity due to chlorosis, reduction of stomatal aperture, regression in shoot biomass, inhibition of root growth also occur under Al stress. Although Al toxicity symptoms are also apparent in the shoots, the roots are the first affected organ. Several researchers indicated that Al-induced root growth inhibition causes poor uptake of soil water and mineral elements such as Ca, Mg, K, Fe and P resulting in decreased yield and crop quality (Jones and Kochian, 1995; Miyasaka et al., 2001; Delhaize et al., 2004).

Several studies prove that root apex (consisted of root cap, meristem and elongation zone) as a first target plays a critical role in Al penetration. This root region also absorbs more Al (Matsumoto, 2000). As a result of Al uptake morphology of root apices change to swollen, cracked, stubby and stiff appearance (Vardar et al., 2006). Reduction in fine branching and root hairs are also evident (Ciamporova, 2002). Root growth inhibition is the most noticeable and evident characteristics of Al toxicity (Singh et al., 2017). It has been considered that reduction of root growth is the consequence of interfered cell elongation and expansion, prior to arrested cell division (Frantzios et al., 2001; Ciamporova, 2002; Vardar et al., 2007). Several researches revealed that plant genotype, cell/tissue type, pH and content of external solution (types of chelators, concentrations of Ca^{2+} and other cations) are decisive on the severity of Al toxicity (Kinraide and Parker, 1987). Table 1 presents some sample of Al toxicity researches performed in different plant species and cell types with different concentration and exposure time.

Al ions interact with cell wall (apoplastic target), plasma membrane and cytosol (symplastic target) within seconds to minutes resulting toxic symptoms (Kochian et al., 2005). It has been identified that Al binds and accumulates in root apoplasm predominantly. Positive charged Al has the ability of binding negatively charged carboxyl groups of pectin building cross-links in cell wall (Blamey et al., 1993; Singh et al., 2017). Al also binds to apoplasmic side of plasma membrane. These reactions cause to the displacement of other positively charged ions such as Ca^{2+} responsible for strength of cell-wall. Al toxicity also induces rapid callose (β-1,3-glucane) synthesis and accumulation on plasma membrane integrating into apoplasm (Tabuchi and Matsumoto, 2001). Callose synthesis which is one of the important markers of Al toxicity may assists plant to manage with Al toxicity by blocking the plasmodesmata. Thus, callose plugs block movement of water and solutes through cell to the apoplasm causing reduction of nutrient supply (Blamey, 2001). Alterations of structural and mechanical functions of cell wall, as a result of strong and rapid binding of Al, cause more rigid and reduced mechanical extensibility. Eventually, disrupted growth orientation together with unusual periclinal divisions occurs (Kochian et al., 2005).

Because of its negatively charged surface, plasma-membrane is one of the important targets for Al^{3+} (Kinraide et al., 1998). It has been reported that Al^{3+} binds to both phospholipids and proteins of the membrane altering the structure by inducing lipid peroxidation (Panda et al., 2009). Al has approximately 560-fold more affinity than other positively charged ions (e.g Ca^{2+}) for the choline head of phosphatidylcholine. Al forms positively charged bridges between head groups of the phospholipid in the plasma membrane resulting in displacement of other cations (Akeson and Munns, 1990; Singh et al., 2017). The positively charged layer changes the movement of ions through the plasma membrane. This situation commonly results in inhibited cation movement and increment in anion movement (Nichol et al., 1993). Consequently, the structure and charge of plasma membrane, phospholipid fluidity altered leading to perturbation of cellular homeostasis (Kochian et al., 2005). It has been also suggested that Al-induced cation displacement triggers callose synthesis (Gupta et al., 2013). Although the Al-specific carrier protein has to be discovered yet, it can interact with different channel proteins inhibiting the uptake of essential ions such as Ca^{2+}, K^{+}, Mg^{2+} and NH^{4+}. Moreover, Al significantly disrupts H^{+} gradient altering the ionic condition and balance of root cells (Piňeros and Tester, 1997; Kochian et al., 2005; Singh et al., 2017).

The cytoskeleton (microtubules, actin filaments and intermediated filaments) is one of the potential targets for Al in the cell causing damage in cytoskeletal dynamics. Cytoskeletal stabilization is important during cell-wall biosynthesis and cell division, principally. Al disrupts cytoskeletal dynamics and stabilization via directly interacting cytoskeletal elements/subunits or altering their signaling cascades (Sivaguru et al., 1999). Researches also revealed that lateral cell swelling is the result of microtubule and actin filament disruption in their organization after Al exposure (Frantzios et al., 2001; Sivaguru et al., 2003). Moreover, Al changes the signal transduction pathways affecting the whole plant. Protein phosphorylation and dephosphorylation are also regulated by Al induction (Matsumoto, 2000; Singh et al., 2017).

Al has a genotoxic profile and long term Al exposure can lead detrimental effects on DNA composition and DNA replication by constructing more rigid double helix and chromatin structure. Many researchers observed that Al toxicity reduces mitotic index and increases abnormal chromosome movements which are related to Al-induced alterations in tubulin polymerization and depolymerization. Besides, chromosome laggards, bridges, micronuclei and c-mitosis are revealed during Al stress (Silva et al., 2000; Frantzios et al., 2000; Vardar et al., 2011).

Al toxicity increases the production and accumulation of reactive oxygen species (ROS) in mitochondria, chloroplast and peroxisome (Kochian et al., 2004) which trigger oxidative damage in cell (Darko et al., 2004; Sharma and Dubey, 2007; Gupta et al., 2013). Under stress conditions various forms of ROS species ($O_2^{\cdot-}$, $^{\cdot}OH$, $HO_2^{\cdot}$, H_2O_2 and $^1O_2$) are generated in plant cell inducing lipid peroxidation, DNA damage, protein denaturation, distribution of antioxidant enzyme balance, carbohydrate oxidation, pigment breakdown leading to programmed cell death.

Table 1. Some sample researches concerning Al toxicity conditions

Plant material	Cell-organ	Growth solution	pH	Al concentration	Exposure time	Reference
Chamelaucium uncinatum	Pollen	Germination medium with boric acid	4.5-6.0	10, 20 µM $AlCl_3$	2 h	Zhang et al., 1999
Hordeum vulgare cv CK218	Root	Nutrient solution	3.5	0.1, 1.0, 10, 30, 50 mM AlCl3	8 h	Pan et al., 2001
Nicotiana tabacum cv Karabağlar	Pollen	Germination medium with boric acid	Unspecified	50, 100, 150, 200 µM $AlCl_3$	24 h	Tuna et al., 2002
N. tabacum cv Samsun	Root	MS medium	4.5	50, 100, 150, 200 µM $AlCl_3$	16 days	Vardar et al., 2006
Zea mays cv Karadeniz Yıldızı	Root	Pretreatment with $CaCl_2$	4.5	150, 300, 450 µM $AlCl_3$	24-48-72-96 h	Vardar et al., 2011
Gossypium hirsutum cv *Nazilli 84S*	Leaves	Soil	4.5	100, 150, 200 µM $AlCl_3$	3 months	Ozyigit et al., 2013
Triticum aestivum, Triticale wittmack, Secale cereale	Root	Direct exposure with Al	4.5	100 µM $AlCl_3$	15-30-60-120 min	Vardar et al., 2015
Solanum lycopersicum cv esculentum and cv cerasiforme	Root and shoot	Clay and sandy soil	Unspecified	280, 560, 1120, 2240 mg/kg Al	18 days	Nogueirol et al., 2015

Plant material	Cell-organ	Growth solution	pH	Al concentration	Exposure time	Reference
Arachis hypogaea cv Zhonghua 2' and cv. '99-1507'	Root	Direct exposure with Al	4.2–4.5	100 µM AlCl3	4-8-12 h	Yao et al., 2016
Citrus sinensis cv Xuegan'and *C. grandis* cv sour pummelo	Leaves	Nutrient solution	4.1-4.2	1.2 mM $AlCl_3$	18 weeks	Li et al., 2016
T. aestivum, H. vulgare, T. wittmack, S. cereale, Avena sativa, Z. mays	Root	Direct exposure with Al	4.5	100 µM $AlCl_3$	0-1/2-1-2-3-4-5-6-7-8 h	Vardar et al., 2016
H. vulgare cv Sebastian	Root	In Hoagland's nutrient solution	4.0	20, 30, 40 µM AlCl3	48-96-144 h	Jaskowiak et al., 2018
N. tabacum cv BY-2	Root and cell culture	Hydrophonic culture and MS medium	4.5	50-100 µM $AlCl_3$	3 days	Kariya et al. 2018

Aluminum Induced Programmed Cell Death

In natural habitats, environmental stresses affect plants frequently because of being sessile organism. Excessive stress conditions, can damage wild flora and agricultural crops with regard to environmental and economic losses worldwide (Tuzhikov et al., 2011). Otherwise, plants have evolved alternative adaptive mechanisms to enable the survival of whole organism. Programmed cell death (PCD) is one of the adaptive mechanisms under extreme environmental stresses (Jackson and Armstrong, 1999; Drew et al., 2000). PCD can be described simply as a genetically regulated and coordinated cell elimination process in certain cells by specific proteases and nucleases during development and biotic–abiotic stresses (Wang et al., 2011; Vardar and Ünal, 2012; Wituszynska and Karpinski, 2013; Petrov et al., 2015).

PCD as a well-organized cell suicide process has common characteristics in eukaryotic cells such as shrinkage of cell, vacuolization, cytochrome c release from mitochondria to cytoplasm, specific protease activation, chromatin condensation, DNA fragmentation and finally breakdown of cellular contents (Qiao et al., 2003; Brighigna et al., 2006; Vardar and Ünal, 2008; Papini et al., 2011; Wang et al., 2012; Poor et al., 2013). Comparing plants and animals, plants have evolved their own pathways due to the existence of cell walls as a barrier. In dying plant cells, the cell wall prevents them from being phagocytosed by neighboring cells as it is in animals. Furthermore, among eukaryotic PCD forms rare conserved regulatory proteins or protein domains have been identified at the molecular level (Danon et al., 2004).

Plant growth and development are mediated and regulated by endogenous factors under environmental stress. One of these endogenous factors is ROS acting as signaling molecules (Nath and Lu, 2015). The ROS are generated as byproducts of respiration and photosynthesis in cellular organelles such as chloroplasts, mitochondria and peroxisomes. Superoxide anion ($O2^{\cdot -}$), hydroxyl radical ($^{\cdot}OH$), hydrogen peroxide (H_2O_2) and singlet oxygen ($^1O_2$) are the most common produced free and non-radicals (Sharma et al., 2012). Lower doses of ROS act as a secondary signaling molecule that mediates the responses toward stress. But higher concentrations of ROS induce oxidative stress which damage macromolecules such as DNA, RNA, protein and lipid leading to PCD (Woo et al., 2013; Coll et al., 2014; Petrov et al., 2015). Even if there are a lot of researches on PCD, the relation between abiotic stress-induced as well as Al stress-induced PCD is still unclear and need to be investigated in molecular aspects.

There are several studies concerning increase of Al-induced ROS and antioxidant enzyme activities which are their scavengers in different plant species. In general, lipid peroxidation, superoxide dismutase, catalase, glutathione and guaiacol peroxidase activities increase in dose and/or time dependent manner (Boscolo et al., 2003; Vardar et al., 2011; Malekzadeh et al., 2015; Nogueirol et al., 2015). Lately, antioxidant enzyme activities and related defense gene expression in two passion fruit cultivars were also evaluated by Pretti et al. (2018). According to their results Al induced oxidative stress and increased both activity and gene expression of catalase and superoxide dismutase characterized by RT-PCR (semi-quantitative reverse transcription-polymerase chain reaction).

It has been revealed that under stress conditions ROS weaken binding of the cytochrome c (cyt c) to the inner mitochondrial membrane and cause its diffusion to the intermembrane space. At the same time, ROS also provoke the mobilization of Ca^{2+} ions from the endoplasmic reticulum to mitochondria. Ca^{2+} ions act as secondary signaling molecules and induce to reduce the mitochondrial membrane potential ($\Delta\Psi m$). Subsequently the mitochondrial permeability transition pores (MPTP) are opened in the outer mitochondrial membrane permitting cyt c pass into the cytosol (Ott et al., 2007; Williams et al., 2014). The cyt c extrusion triggers more ROS generation in the mitochondria amplifying the signal. According to the researches, enhanced ROS generation and cyt c release trigger caspase-like activities in plants (Gunawardena and McCabe, 2015).

Li and Xing (2011) investigated possible mechanisms of PCD induced by Al. The researchers used to monitor the behavior of mitochondria and increase of caspase-3-like activity. They detected a quick burst of mitochondrial ROS, swelling of mitochondria, loss of MTP (mitochondrial transmembrane potential) and caspase-3-like activation prior to PCD in Arabidopsis. The researchers concluded that mitochondrial events such as swelling, generation of ROS and loss of MTP have critical roles triggering caspase-3-like activity and PCD providing a new perspective about Al-induced PCD mechanism.

Huang et al. (2014) reported that Al-induced ROS accumulation and PCD are closely related. According to their results 100 μM $AlCl_3$ caused ROS burst and up-regulation of *Rboh* and *COX* expression related to ROS mediation in peanut roots. Besides, the researchers also revealed increased MPTP opening, decreased $\Delta\Psi m$, cyt c release from mitochondria to cytoplasm and caspase 3-like protease activity. The researchers concluded that Al induced PCD is regulated via a cascade of cellular events depending on mitochondrial pathway and ROS mechanism. Zhan et al. (2014) undertook a research to establish the role of mitochondria in Al-induced PCD in both Al-tolerant and sensitive peanuts. It has been reported that increase in mitochondrial ROS production and MDA were enhanced in two peanut cultivars. Besides, reduction of mitochondrial Ca concentration, opening of MPTP, collapse of $\Delta\Psi m$ and cyt c release were more extensively in Al sensitive cultivar depending on Al concentration.

In animals, caspases regulate the PCD. Caspases are cysteine-dependent proteases and cleave their substrates after an Asp (aspartic acid) residue (Cai et al., 2014). It has been indicated that there are no caspase homologs in plant genomes, but caspase-like proteases and caspase-like activities were also identified in plants recently (Rocha et al., 2017). Synthetic animal caspase substrates were also used to determine caspase-like activities in plant PCD types (McStay et al., 2007; Ayturk and Vardar, 2015). In last decades, plant metacaspases are accepted as distant relatives of animal caspases (Uren et al., 2000). Metacaspases prefer to cleave their substrates after arginine (Arg) or lysine (Lys) as distinct from animal caspases (Watanabe and Lam, 2005; He et al., 2008; Salvesen et al., 2016). Similar to caspases and metacaspases, vacuolar processing enzymes (VPEs) are also cysteine proteases (Hatsugai et al., 2004; Cai et al., 2014). VPEs cleave their substrates after an asparagine (Asn) residue distinct from caspases and metacaspases, but interestingly they also interact and cleave caspase-1 substrate (YVAD). Moreover, VPE activation repressed by inhibitor of synthetic caspase-1(ac-YVAD-CHO) (Sexton et al., 2007; Misas-Villamil et al., 2013). It has been reported that both metacaspases and VPEs have a regulatory role during environmental stress-induced PCD.

Kariya et al. (2013) investigated the role of vacuole in Al-induced PCD mechanism in *Nicotiana tabacum* L. cv BY-2 cell line. After 18 h Al treatment (0-150 μM), cell growth and the integrity of plasma membrane were reduced, whereas VPE activity was increased dose dependently. Besides, morphological alterations of vacuole were also evident. They also observed that the presence of inhibitor of VPE (Ac-YVAD-CHO) suppressed the loss of plasma membrane integrity. The researchers evidenced that Al-induced VPE activity is controlled at transcriptional level, and loss of integrity of the plasma membrane and a loss of growth capacity are regulated by VPE activity. Lately Kariya et al. (2018) reported an increase in gene expression of *VPE1a* and *VPE1b*, but there were no change in *VPE2* and *VPE3* gene expression at tobacco roots after Al exposure.

Several studies using synthetic caspase substrates revealed caspase-like activities under Al stress conditions. Aytürk and Vardar (2015) indicated caspase-3, -8 and -9 like activities in triticale, barley, rye and oat roots after 100 μM Al exposure. Similarly, Yao et al. (2016) used synthetic caspase-1, -2, -3, -4, -5, -6, -8 and -9 substrates in Al tolerant and Al sensitive peanut roots after 100 μM Al. The results demonstrated that especially caspase-1 and -3 activities showed the highest activity in both cultivars. Besides, Al toxicity caused nuclear degeneration and DNA fragmentation.

DNA fragmentation originates from internucleosomal DNA cleavage by specific proteases and nucleases. Whereas ultrastructural features such as cell shrinkage and vacuolization points PCD (Liu et al., 2018), DNA fragmentation analysis is the most sensitive parameter of PCD. DNA fragmentation can be monitored by using TUNEL (Terminal deoxynucleotidyl transferase dUTP nick end labeling) reaction histochemically, comet assay, flow cytometry and laddering on the agarose gel.

In one of the early studies assessing Al-induced DNA fragmentation, barley roots were exposed to Al within a range of 0.1–50 mM (Pan et al., 2001). It has been revealed that Al concentrations between 0.1–1 mM induced clear DNA fragmentation, but the higher concentrations (10–50 mM) caused necrotic DNA smear. Comet assay can be also used to indicate DNA damage by measuring the DNA tail. Achary et al. (2008) demonstrated Al-induced breaks in double stranded DNA in *Allium cepa* roots by comet assay. Subsequently, Achary and Panda (2009) also evaluated Al-induced DNA damage in a range of 200-800 μM Al in *A. cepa*. Achary et al. (2012) revealed a dose-dependent DNA damage in *H. vulgare* roots with different concentrations of Al (2.5, 5 and 10 mM). Furthermore, in peanut (*Arachis hypogaea*) roots 100 μM $AlCl_3$ induced DNA fragmentation (Huang et al., 2014) revealed time dependent DNA damage by comet assay in wheat, rye and triticale roots after 100 μM $AlCl_3$

exposure. Similarly, Vardar et al. (2016) demonstrated DNA fragmentation by agarose gel electrophoresis in wheat, oat, rye and triticale. Otherwise, Jaskowiak et al. (2018) reported Al-induced DNA fragmentation by TUNEL reaction in barley roots. The researchers also tested cell cycle profile and DNA integrity after Al exposure by flow cytometry. According to the results, Al reduced the frequency of S phase cells and altered the cell cycle profile significantly. The delay of mitotic divisions caused increased cells frequency in G2/M phase.

Although there are several researches on Al-induced DNA damage, Al induced DNA methylation and polymorphism on LTR retrotransposons were also reported recently (Guo et al., 2018; Taşpınar et al., 2018). The researchers suggested that transposon polymorphism and methylation may be a part of the defense mechanism under Al stress.

Aluminum Tolerance and Alleviation

It has been known that some plant genotypes are Al tolerant. Because of the agronomic importance of Al resistance, the underlying molecular and physiological principles have become an important research area. The mechanisms of Al tolerance can be regulated both exclusion of Al from the root apex or intracellular tolerance of Al being transported into the symplast (Kochian et al., 2005).

It has been known that Al as a very reactive cation tends to form strong complexes with oxygen donor ligands. Several researches revealed that root exudates or mucilage (glucose, galactose, arabinose and uronic acids) make complexation with Al via chelation or direct binding preventing the accumulation of Al in apoplast or symplast (Barceló and Poschenrieder, 2002). Root apex secretes phosphate, carboxylate, phenolic substances, pectates, mucopolysaccharides and/or siderophores which can make complexes with Al ions (Winkler et al., 1986). Chelation of Al blocks the activity of Al ions preventing their binding to Al targets such as cell wall and plasma membrane. Although roots secrete different variety of exudates, carboxylates such as malate and citrate are most commonly detected (Barceló and Poschenrieder, 2002).

Whereas root exudates seem to be the most important Al tolerance mechanism restricting Al transport to the upper organs, several plant genotypes may tolerate high Al levels in their cell vacuoles or apoplast with ligands in an innocuous form by complexation and detoxification of Al. For example, *Hydrangea macrophylla* leaves accumulate Al making complex with citric acid and their sepals turn red to blue due to accumulated Al concentration. This situation also related to soil acidification resulting in Al dissolution (Takeda et al., 1985). The stated internal detoxification mechanism occurs with Al chelation in the cytosol and storage in the vacuole as Al-carboxylate complex.

Recently several types of ameliorating chemicals against Al toxicity have attracted attention in breeding crops because of their agronomic importance. Based on this, many researchers experienced some chemicals exogenously which are tend to make strong complexes with Al (Table 2). The most prominent chemicals are citric, oxalic, malic, oxalic and tartaric acids which are oxygen donor ligands (Hue et al. 1986; An et al., 2014). Yalçın and Vardar (2016) reported that both pre-treatment and co-treatment of salicylic acid alleviated Al toxicity, possibly stimulating the roots to secrete organic molecules such as citrate which are able to make chelation with Al. Pandey et al. (2013) alleviated Al toxicity applying Ca, Mg and salicylic acid. According to their results, the exogenously supplied chemicals suppressed ROS mediated oxidative stress, but salicylic acid was more effective in alleviation than Ca and Mg. Besides Chen et al. (2013) investigated the possible role of H_2S against Al toxicity. The researchers revealed that H_2S reduced Al accumulation and ROS generation by increased antioxidant enzyme activities. They also indicated that H_2S increased both citrate secretion and citrate transporter gene expression. Wang et al. (2013) determined the alleviation effect of indol-3-acetic acid through reducing Al uptake and stimulating malic acid secretion in wheat roots. Moreover, it has been reported that silicone reduces Al uptake and ameliorates Al toxicity by making complexes with Al ions

(Kopittke et al., 2017). It has been known that polyamine (putrescine, spermidine, spermine) accumulation has a correlation with stress tolerance and adaptation (Nahar et al., 2017). Based on this, Yu et al. (2016) indicated a different mechanism on the alleviation of Al toxicity by using putrescine. The researchers revealed that putrescine induced on ethylene synthesis mechanism and decreased ethylene production alleviating root growth inhibition under Al stress conditions. Similarly, Nahar et al. (2017) reported alleviating effects of spermidine in mung bean seedlings under Al stress. The researchers also suggested that polyamines might have roles in regulating the antioxidant enzyme activities to decrease oxidative stress during Al stress.

In conclusion; Al toxicity especially in acidic soils has adverse effects on plant growth and development, as a result of this the productivity of crop reduces. Although the toxic effects of Al are well documented, the molecular signaling mechanism of Al toxicity is not clear. Besides, there is limited studies on Al-induced PCD and needs more detailed and sensitive research. To identify the Al toxicity and tolerance mechanism may help to improve Al tolerant varieties. Although several researches performed to mitigate Al toxicity using several types of chemicals, their alleviation mechanisms are still need to be clarified. Beside the Al tolerant varieties, proven chemicals might be used to ameliorate the adverse effects of Al to enhance the crop quality.

Table 2. Some ameliorating chemicals against to Al toxicity

Plant material	Plant organ	Growth solution	Al conc.	Alleviation agent	Reference
Oryza sativa	Root, shoot	Direct exposure	0.5 mM $AlCl_3$	1 mM $CaCl_2$ 0.25 mM $MgSO_4$ 60 μM Salicylic acid	Pandey et al., 2013
Triticum aestivum	Root	$CaCl_2$	50 $AlCl_3$	25 μM Indol acetic acid 5 μM Niflumic acid 5 μM 2,3,5-triiodobenzoic acid	Wang et al., 2013
Medicago sativa	Leaves	Hoagland solution (spray)	100 μM $AlCl_3$	100 μM Succinic acid 100 μM Citric acid 100 μM Malic acid 100 μM Oxalic acid	An et al., 2014

Plant material	Plant organ	Growth solution	Al conc.	Alleviation agent	Reference
T. aestivum	Root	Nutrient solution	100 μM $AlCl_3$	125- 625-2500 μM $CaCl_2$	Hossain et al., 2014
Brassica napus	Shoot, root, leaves	Hoagland solution	0.1-0.2-0.3-0.6-1 mM $AlCl_3$	0.3 mM NaHS	Ali et al., 2014
Citrus grandis	Shoot, root	Nutrient solution	1.2 mM $AlCl_3$	2,5-20 μM H_3BO_3	Zhou et al., 2015
T. aestivum	Shoot, root	Oxisol slury	-	Silicon (Si) rich biochar	Qian et al., 2016
Hordeum vulgare	Root	Direct exposure	20 μM $AlCl_3$	5 -10 μM Salicylic acid	Yalcın and Vardar, 2016
Borago officinalis	Root, shoot	Hydrophonic solution	20-60 mM $AlCl_3$	0.2-1-2 mM Si	Delavar et al., 2016
Glycine max, Sorghum bicolor	Root	Direct exposure	5-10-25-30 μM $AlCl_3$	2 mM Si	Kopittke et al., 2017
O. sativa	Root, shoot	Nutrient solution	0.19 mM Al	0.11 mM Si	Freitas et al., 2017
O. sativa	Root	$CaCl_2$ solution	50 μM Al	10 μM La 400-600 μL/L CO_2	Zhu et al., 2017
Vigna radiata	Leaves	Direct exposure	0.5 mM $AlCl_3$	0.3 mM spermidine	Nahar et al., 2017
H. vulgare	Root, shoot	Hoagland solution	2-4-6 mM	2 mM Ascorbic acid 250 μM Salicylic acid	Shahnawaz et al., 2017
Trifoliate orange	Root	Direct exposure	400 μM $AlCl_3$	10 μM H_3BO_3	Riaz et al., 2018
O. sativa	Root	$CaCl_2$ solution	30-50-100 μM $AlCl_3$	1-2-5-10 μM NaHS	Zhu et al., 2018
T. aestivum	Root	$CaCl_2$ solution	30 μM $AlCl_3$	0.1-0.3-0.5 mM glucose oxidase 100 μM NO scavenger 100 μM diphenylene iodonium 5 mM dimethylthiourea 100 μM nitrate reductase 100 μM nitric oxide synthase	Sun et al., 2018

ACKNOWLEDGMENT

This review is performed by the support of Research Foundation of Marmara University (BAPKO no. FEN-A-070617-0360).

REFERENCES

Abate, E; Hussien, S; Laing, M; Mengistu, F. (2013). Aluminum toxicity tolerance in cereals: Mechanisms, genetic control and breeding methods. *African Journal of Agricultural Research,* 8, 711-722.

Achary, VMM; Panda, BB. (2009). Aluminium-induced DNA damage and adaptive response to genotoxic stress in plant cells are mediated through reactive oxygen intermediates. *Mutagenesis*, 25, 201-209.

Achary, VMM; Jena S; Panda KK; Panda BB. (2008). Aluminium induced oxidative stress and DNA damage in root cells of *Allium cepa* L. *Ecotoxicology and Environmental Safety*, 70, 300-310.

Achary, VMM; Patnaik, AR; Panda, BB. (2012). Oxidative biomarkers in leaf tissue of barley seedlings in response to aluminium stress. *Ecotoxicology and Environmental Safety*, 75, 16-26.

Akeson, MA; Munns, DN. (1990). Uptake of aluminum into root cytoplasm; predicted rates for important solution complexes. *Journal of Plant Nutrition*, 13, 467-484.

Ali, B; Qian P; Sun, R; Farooq, MA; Gill, RA; Wang, J; Azam, M; Zhou W. (2014).Hydrogen sulfide alleviates the aluminum-induced changes in *Brassica napus* as revealed by physiochemical and ultrastructural study of plant. *Environmental Science and Pollution Research*, 22, 3068-81

An, Y; Zhou, P; Xiao, Q; Shi, D. (2014). Effects of foliar application of organic acids on alleviation of aluminum toxicity in alfalfa. *Journal of Plant Nutrition Soil Science*, 177, 421-430.

Aytürk, Ö; Vardar, F. (2015). Aluminum induced caspase-like activities in some Gramineae species. *Advances in Food Sciences*, 37, 71-75.

Barceló, J; Poschenrieder, C. (2002). Fast root growth responses, root exudates and internal detoxification as clues to the mechanisms of aluminum toxicity and resistance: A review. *Environmental and Experimental Botany,* 48, 75-92.

Blamey, FPC; Asher, CJ; Kerven, GL; Edwards, DG. (1993). Factors affecting aluminum sorption by calcium pectate. *Plant and Soil*, 149, 87-94.

Blamey, FPC. (2001). The role of the root cell wall in aluminum toxicity. In: Ae, N; Arihara, J; Okada, K; Srinivasan, A. *Plant Nutrient Acquisition, New Perspectives.* Springer Verlag, New York, 201-226.

Boscolo, PRS; Menossi, M; Jorge, RA. (2003). Aluminum-induced oxidative stress in maize. *Phytochemistry*, 62, 181-189.

Brighigna, L; Milocani, E; Papini, A; Vesprini, JL. (2006). Programmed cell death in the nucellus of *Tillandsia* (Bromeliaceae). *Caryologia*, 59, 334-339.

Cai, YM; Yu, J; Gallois, P. (2014). Endoplasmic reticulum stress-induced PCD and caspase-like activities involved. *Frontiers in Plant Science*, 5, 41.

Chen, J; Wang, WH; Wu, FH; You, CY; Liu, TW; Dong, XJ. (2013). Hydrogen sulfide alleviates aluminum toxicity in barley seedlings. *Plant Soil*, 362, 301-318.

Ciamporova, M. (2002). Morphological and structural responses of plant roots to aluminum at organ, tissue and cellular levels. *Biologia Plantarum,* 45, 161-171.

Coll, NS; Smidler, A; Puigvert, M; Popa, C; Valls, M; Dangl, JL. (2014). The plant metacaspase AtMC1 in pathogen-triggered programmed cell death and aging: functional linkage with autophagy. *Cell Death and Differentiation*, 21, 1399.

Danon, A; Rotari, VI; Gordon, A; Mailhac, N; Gallois, P. (2004). Ultraviolet-C overexposure induces programmed cell death in *Arabidopsis* which is mediated by caspase-like activities and which can be suppressed by caspase inhibitors, p35 and defender against apoptotic death. *Journal of Biological Chemistry*, 279, 779-787.

Darko, E; Ambrus, H; Stefanovits-Bányai, E; Fodor, J; Bakos, F; Barnabás, B. (2004). Aluminium toxicity, Al tolerance and oxidative stress in an Al-sensitive wheat genotype and in Al-tolerant lines developed by in vitro microspore selection. *Plant Science*, 166, 583-591.

Delavar, K; Enteshari, S; Gagoonani, S; Kamali, S. (2016). Effect of silicon on the mineral content of borage (*Borago officinalis* L.) under aluminum stress. *Journal of Biological Sciences*, 16, 128-135.

Delhaize, E; Ryan, PR; Hebb, DM; Yamamoto, Y; Sasaki, T; Matsumoto, H. (2004). Engineering high-level aluminum tolerance in barley with the ALMT1 gene. *PNAS*, 101, 15249-54.

Drew, MC; He, CJ; Morgan, PW. (2000). Programmed cell death and aerenchyma formation in roots. *Trends in Plant Science*, 5, 123-127.

Eswaran, H; Reich, P; Beinroth, F. (1997). Global distribution of soils with acidity. In: Moniz, AC. *Plant-Soil Interactions at low pH*. Brazilian Soil Science Society, Brazil, 159-164.

Frantzios, G; Galatis, B; Apostolakos, P. (2000). Aluminum effects on microtubule organization in dividing root tip cells of *Triticum turgidum*. I. Mitotic cells. *New Phytology*, 145, 211-224.

Frantzios, G; Galatis, B; Apostolakos P. (2001). Aluminum effects on microtubule organization in dividing root tip cells of *Triticum turgidum*. II. Cytokinetic cells. *Journal of Plant Research*, 114, 157–170.

Freitas, LB; Fernandes, DM; Maia, SCM; Fernandes, AM. (2017). Effects of silicon on aluminum toxicity in upland rice plants. *Plant Soil*, 420, 263-275.

Gunawardena, AH; McCabe, PF. (2015). *Plant Programmed Cell Death*. Springer.

Guo, P; Qi, YP; Cai, YT; Yang, TY; Yang, LT; Huang, ZR; Chen, LS. (2018). Aluminum effects on photosynthesis, reactive oxygen species and methylglyoxal detoxification in two *Citrus* species differing in aluminum tolerance. *Tree Physiology*, 00, 1-18.

Gupta, N; Gaurav, SS; Kumar, A. (2013). Molecular basis of aluminium toxicity in plants: A review. *American Journal of Plant Science*, 4, 21-37.

Hatsugai, N; Kuroyanagi, M; Yamada, K; Meshi, T; Tsuda, S; Kondo, M; Nishimura, M; Hara-Nishimura, I. (2004). A plant vacuolar protease, VPE, mediates virus-induced hypersensitive cell death. *Science*, 305, 855-858.

He, R; Drury, GE; Rotari, VI; Gordon, A; Willer, M; Farzaneh, T; Woltering, EJ; Gallois, P. (2008). Metacaspase-8 modulates programmed cell death induced by ultraviolet light and H_2O_2 in *Arabidopsis*. *Journal of Biological Chemistry*, 283, 774-783.

Hossain, MA; Ashrafuzzaman, M; Hossain, AKMZ; Ismail, MR; Koyama, H. (2014). Role of accumulated calcium in alleviating aluminum injury in wheat plants. *The Scientific World Journal*, ID 457187.

Huang, WJ; Oo, TL; He, HY; Wang, AQ; Zhan, J; Li, CZ; Wei, SQ; He, LF. (2014). Aluminum induced rapidly mitochondria dependent programmed cell death in Al sensitive peanut root tips. *Botanical Studies*, 55, 67-78.

Hue, NV; Craddock, GR; Adams, F. (1986). Effect of organic acids on aluminum toxicity in subsoils. *Soil Science Society of America Journal Abstract*, 50, 28-34.

Jackson, MB; Armstrong, W. (1999). Formation of aerenchyma and the processes of plant ventilation in relation to soil flooding and submergence. *Plant Biology*, 1, 274-287.

Jaskowiak, J; Tkaczyk, O; Slota, M; Kwasniewska1, J; Szarejko, I. (2018). Analysis of aluminum toxicity in *Hordeum vulgare* roots with an emphasis on DNA integrity and cell cycle. *PLOS ONE*, 13, e0193156.

Jones, DL; Kochian, LV. (1995). Aluminum inhibition of the inositol 1,4,5-triphosphate signal transduction pathway in wheat roots – a role in aluminum toxicity. *Plant Cell*, 7, 1913-1922.

Kariya, K; Demiral, T; Sasaki, T; Tsuchiya, Y; Turkan, I; Sano, T; Yamamoto, Y. (2013). A novel mechanism of aluminium-induced cell death involving vacuolar processing enzyme and vacuolar collapse in tobacco cell line BY-2. *Journal of Inorganic Biochemistry*, 128, 196-201.

Kariya, K; Tsuchiya, Y; Sasaki, T; Yamamoto, Y. (2018). Aluminium-induced cell death requires upregulation of NtVPE1 gene coding vacuolar processing enzyme in tobacco (*Nicotiana tabacum* L.). *Journal of Inorganic Biochemistry*, 181, 152-161.

Kinraide, TB; Parker, DR. (1987). Cation amelioration of aluminum toxicity in wheat. *Plant Physiology*, 83, 546-551.

Kinraide, TB; Yermiyahu, U; Rytwo, G. (1998). Computation of surface electrical potentials of plant cell membranes. Correspondence to published zeta potentials from diverse plant sources. *Plant Physiology*, 118, 505-512.

Kochian, LV; Hoekenga, OA; Pineros, MA. (2004). How do crop plants tolerate acid soils? Mechanisms of aluminum tolerance and phosphorous efficiency. *Annual Review of Plant Biology*, 55, 459-493.

Kochian, LV; Piñeros, MA; Hoekenga, OA. (2005). The physiology, genetics and molecular biology of plant aluminum resistance and toxicity. *Plant and Soil*, 274, 175-195.

Kopittke, PM; Gianoncelli, A; Kourousias, G; Green K; McKenna, BA. (2017). Alleviation of Al toxicity by Si is associated with the formation of Al–Si complexes in root tissues of *Sorghum*. *Frontiers in Plant Science*, 8, 2189.

Li, H; Yang, LT; Qi, YP; Guo, P; Lu, YB; Chen, LS. (2016). Aluminum toxicity-induced alterations of leaf proteome in two citrus species differing in aluminum tolerance. *International Journal of Molecular Sciences*, 17, 1180.

Li, Z; Xing, D. (2011). Mechanistic study of mitochondria-dependent programmed cell death induced by aluminium phytotoxicity using fluorescence techniques. *Journal of Experimental Botany*, 62, 331–343.

Liu, W; Xu, F; Lv, T; Zhou, W; Chen, Y; Jin, C; Lu, L; Lin X. (2018). Spatial responses of antioxidative system to aluminum stress in roots of wheat (*Triticum aestivum* L.) plants. *Science of the Total Environment*, 627, 462-469.

Ma, JF; Chen, ZC; Shen, RF. (2014). Molecular mechanism of Al tolerance in Gramineous plants. *Plant and Soil*, 381, 1-12.

Malekzadeh, P; Mehr, RS; Hatamnia, AA. (2015). Effects of aluminum toxicity on maize (*Zea mays* L.) seedlings. *Iranian Journal of Plant Physiology,* 5, 1289-1296.

Matsumoto H. (2000). Cell biology of aluminum toxicity and tolerance in higher plants. I*nternational Review of Cytology,* 200, 1-47.

McStay, GP; Salvesen, GS; Green, DR. (2007). Overlapping cleavage motif selectivity of caspases: implications for analysis of apoptotic pathways. *Cell Death and Differentiation*, 15, 322-331.

Misas-Villamil, JC; Toenges, G; Kolodziejek, I; Sadaghiani, AM; Kaschani, F; Colby, T; Bogyo, M; van der Hoorn, RAI. (2013). Activity profiling of vacuolar processing enzymes reveals a role for VPE during oomycete infection. *Plant Journal*, 73, 689-700.

Miyasaka, SC; Hawes, C. (2001). Possible role of root border cells in detection and avoidance of aluminum toxicity. *Plant Physiology*, 125, 1978-1987.

Nahar, K; Hasanuzzaman, M; Suzuki T; Fujita, M. (2017). Polyamines-induced aluminum tolerance in mung bean: A study on antioxidant defense and methylglyoxal detoxification systems. *Ecotoxicology*, 26, 58-73.

Nath, K; Lu, Y. (2015). A Paradigm of reactive oxygen species and programmed cell death in plants. *Journal of Cell Science and Therapy*, 6.

Nichol, B; Oliveria, LA; Glass, ADM; Siddiqi, MY. (1993). The effects of aluminum on the influx of calcium, potassium, ammonium, nitrate and phosphate in an aluminum-sensitive cultivar of barley (*Hordeum vulgare* L.). *Plant Physiology*, 101, 1263-1266.

Nogueirol, RC; Monteiro, FA; Gratão PL; Borgo L; Azevedo RA. (2015). Tropical soils with high aluminum concentrations cause oxidative stress in two tomato genotypes. *Environmental Monitoring and Assesment*, 187, 73.

Ozyigit, II; Vardar, F; Yaşar, U; Akıncı, Ş. (2013). Long term effects of aluminum and cadmium on growth, leaf anatomy, and photosynthetic pigments of cotton. *Communication in Soil Science and Plant Analysis*, 44, 3076-3091.

Ott, M; Gogvadze, V; Orrenius, S; Zhivotovsky, B. (2007). Mitochondria, oxidative stress, and cell death. *Apoptosis*, 12, 913-922.

Pan, JW; Zhu, MY; Chen H. (2001). Aluminum-induced cell death in root-tip cells of barley. *Environmental and Experimental Botany*, 46, 71-79.

Panda, SK; Baluška, F; Matsumoto, H. (2009). Aluminum stress signaling in plants. *Plant Signaling and Behavior*, 4, 592-597.

Pandey, P; Srivastava, RK; Dubey, RS. (2013). Salicylic acid alleviates aluminum toxicity in rice seedlings better the magnesium and calcium by reducing aluminum uptake, suppressing oxidative damage and increasing antioxidative defense. *Ecotoxicology*, 22, 656-670.

Papini, A; Mosti, S; Milocani, E; Tani, G; Di Falco, P; Brighigna, L. (2011). Megasporogenesis and programmed cell death in *Tillandsia* (Bromeliaceae). *Protoplasma*, 248; 651-662.

Petrov, V; Hille, J; Mueller-Roeber, B; Gechev, TS. (2015). ROS mediated abiotic stress-induced programmed cell death in plants. *Frontiers in Plant Science*, 6, 69.

Piňeros, MA; Tester, M. (1997). Calcium channels in higher plant cells: Selectivity, regulation and pharmacology. *Journal of Experimental Botany*, 48, 551-577.

Poor, P; Kovacs, J; Szopko, D; Tari I. (2013). Ethylene signaling in salt stress- and salicylic acid-induced programmed cell death in tomato suspension cells. *Protoplasma*, 250, 273-284.

Pretti, IR; Luz, AC; Perdigão, TL; Araújo RA;Batitucci, MCP. (2018). Gene expression and antioxidant enzymatic activity in passion fruit exposed to aluminum. *African Journal of Agricultural Research*, 13, 115-120.

Qian, L; Chen, B; Chen, M. (2016). Novel alleviation mechanisms of aluminum phytotoxicity via released biosilicon from rice straw-derived biochars. *Nature Scientific Reports*, 6, 29346.

Qiao, JJ; Yuan, YJ; Zhao, H; Wu, JC. (2003). Differences in protein patterns in suspension cultures of *Taxus cuspidata* induced by cerium. *Biology Plantarum*, 46, 611-615.

Riaz, M; Yan, L; Wu, X; Hussain, S; Aziz, O; Wang, Y; Imran, M; Jiang, C. (2018). Boron alleviates the aluminum toxicity in trifoliate orange by regulating antioxidant defense system and reducing root cell injury. *Journal of Environmental Management*, 208, 148-158.

Rocha, GL; Fernandez, JH; Oliveira, AEA; Fernandes, KVS. (2017). *Programmed cell death-related proteases in plants*. In Tech, Croatica.

Salvesen, GS; Hempel, A; Coll, NS. (2016). Protease signaling in animal and plant-regulated cell death. *FEBS Journal*, 283, 2577-2598.

Sexton, KB; Witte, MD; Blum, G; Bogyo, M. (2007). Design of cell-permeable fluorescent activity-based probes for the lysosomal cysteine protease asparaginylendopeptidase (AEP)/legumain. *Bioorganic and Medicinal Chemistry Letters*, 17, 649-653.

Shahnawaz, MD; Chouhan, R; Sanadhya, D. (2017). Impact of aluminum toxicity on physiological aspects of barley (*Hordeum vulgare* L) cultivars and its alleviation through ascorbic acid and salicylic acid seed priming. *International Journal of Current Microbiology and Applied Sciences*, 6, 875-891.

Sharma, P; Jha, AB; Dubey, RS; Pessarakli, M. (2012). Reactive oxygen species, oxidative damage, and antioxidative defense mechanism in plants under stressful conditions. *Journal of Botany*, 217037.

Sharma, P; Dubey, RS. (2007). Involvement of oxidative stress and role of antioxidative defense system in growing rice seedlings exposed to toxic concentrations of aluminum. *Plant Cell Reports*, 26, 2027-2038.

Silva, I; Smyth, T; Moxley, D; Carter, T; Allen, N; Rufty, T. (2000). Aluminum accumulation at nuclei of cells in the root tip. Fluorescence detection using lumogallion and confocal laser scanning microscopy. *Plant Physiology*, 123, 543-552.

Singh, S; Tripathi, DK; Singh, S; Sharma, S; Dubey, NK; Chauhan, DK; Vaculík M. (2017). Toxicity of aluminium on various levels of plant cells and organism: A review. *Environmental and Experimental Botany*, 137, 177-193.

Sivaguru, M; Baluska, F; Volkmann, D; Felle, HH; Horst, WJ. (1999). Impacts of aluminum on the cytoskeleton of the maize root apex. Short term effects on the distal part of the transition zone. *Plant Physiology*, 119, 1073-1082.

Sivaguru, M; Pike, S; Gassmann, W; Baskin, T. (2003). Aluminum rapidly depolymerizes cortical microtubules and depolarizes the plasma membrane: Evidence that these responses are mediated by a glutamate receptor. *Plant Cell Physiology,* 44, 667–675.

Sun, C; Liu, L; Lu, L; Jin, C; Lin, X. (2018). Nitric oxide acts downstream of hydrogen peroxide in regulating aluminum induced antioxidant defense that enhances aluminum resistance in wheat seedlings. *Environmental and Experimental Botany*, 145, 95-103.

Tabuchi, A; Matsumoto, H. (2001). Changes in cell wall properties of wheat (*Triticum aestivum*) roots during aluminum-induced growth inhibition. *Physiology Plantarum,* 112, 353-358.

Takeda, K; Kariuda, M; Itoi, H. (1985). Blueing of sepal colour of *Hydrangea macrophylla*. *Phytochemistry*, 24, 2251-2254.

Taspinar, MS; Aydin, M; Sigmaz, B; Yagci, S; Arslan, E; Agar, G. (2018). Aluminum-induced changes on DNA damage, DNA methylation and LTR retrotransposon polymorphism in maize. *Arabian Journal of Science and Engineering*, 43, 123-131.

Tuna, AL; Bürün, B; Yokaş, İ; Çoban, E. (2002). The effects of heavy metals on pollen germination and pollen tube length in the tobacco plant. *Turkish Journal of Biology*, 26, 109-113.

Tuzhikov, AI; Vartapetian, BB; Vartapetian, AB; Chichkova, NV. (2011). Abiotic stress-induced programmed cell death in plants: A phytaspase connection. In: Shanker, A. *Abiotic stress responses in plants – Physiological, biochemical and genetic perspectives*. Intech, Croatica, 183-196.

Uren, AG; O'Rourke, K; Aravind, L; Pisabarro, MT; Seshagiri, S; Koonin, EV; Dixit, VM. (2000). Identification of paracaspases and metacaspases: Two ancient families of caspase-like proteins, one of which plays a key role in MALT Lymphoma. *Molecular Cell,* 6, 961-967.

Vardar, F; Arıcan, E; Gözükırmızı, N. (2006). Effects of aluminum on *in vitro* root growth and seed germination of tobacco (*Nicotiana tabacum* L.). *Advances in Food Science*, 28, 85-88.

Vardar, F; Ünal, M. (2007). Aluminum toxicity and resistance in higher plants. *Advances in Molecular Biology,* 1, 1-12.

Vardar, F; Ünal, M. (2008). Proteolytic enzymes in plant programmed cell death. *Turkish Journal of Scientific Reviews*, 1, 65-78.

Vardar, F; İsmailoğlu, I; İnan, D; Ünal, M. (2011). Determination of stress responses induced by aluminum in maize (*Zea mays* L. Karadeniz Yıldızı). *Acta Biologica Hungarica*, 62, 156-170.

Vardar, F; Ünal, M. (2012). Ultrastructural aspects and programmed cell death in the tapetal cells of *Lathyrus undulatus* Boiss. *Acta Biologica Hungarica*, 63, 56-70.

Vardar, F; Akgül, N; Aytürk, Ö; Aydın, Y. (2015). Assessment of aluminum induced genotoxicity with comet assay in wheat, rye and triticale roots. *Fresenius Environmental Bulletin*, 37, 3352-3358.

Vardar, F; Çabuk, E; Aytürk, Ö; Aydın, Y. (2016). Determination of aluminum induced programmed cell death characterized by DNA fragmentation in Gramineae species. *Caryologia*, 69, 111-115.

Vitorello, VA; Capaldi, FRC; Stefanuto VA. (2005). Recent advances in aluminum toxicity and resistance in higher plants. *Brazilian Journal of Plant Physiology,* 17, 129-143.

Wang, J; Li, X; Liu, Y; Zhao, X; Chen, C; Tian, F. (2011). MEK/ERK inhibitor U0126 enhanced salt stress-induced programmed cell death in *Thellungiella halophila* suspension-cultured cells. *Plant Growth and Regulation*, 63, 207-216.

Wang, GP; Zhang, ZH; Kong, DJ; Liu, QX; Zhao, GL. (2012). Programmed cell death is responsible for replaceable bud senescence in chestnut (*Castanea mollissima* BL.). *Plant Cell Reports*, 31, 1603-1610.

Wang, Q; Nian, F; Zhao, L; Li, F; Yang, H; Yang, Y. (2013). Exogenous indole-3-acetic acid could reduce the accumulation of aluminum in root apex of wheat (*Triticum aestivum* L.) under Al stress. *Journal of Soil Science and Plant Nutrition*, 13, 534-543.

Watanabe, N; Lam, E. (2005). Two *Arabidopsis* metacaspases AtMCP1b and AtMCP2b are arginine/lysine-specific cysteine proteases and activate apoptosis-like cell death in yeast. *Journal of Biological Chemistry*, 280, 14691-14699.

Williams, B; Verchot, J; Dickman, MB. (2014). When supply does not meet demand-ER stress and plant programmed cell death. *Frontiers in Plant Science*, 5, 211.

Winkler, S; Ockels, W; Budzikiewicz, H; Korth, H; Pulverer, G. (1986). 2-hydroxy-4-methoxy-5-methylpyridine N-oxide: an aluminum complexing metabolite from *Pseudomonas cepacia. Zeitschrift für Naturforschung*, 41, 807-808.

Wituszyńska, W; Karpiński, S. (2013).Programmed cell death as a response to high light, UV and drought stress in plants. In: Vahdati, K; Leslie, C; *Abiotic stress plant responses and applications in agriculture*. InTech, Croatica, 207-246.

Woo, HR; Kim, HJ; Nam, HG; Lim, PO. Plant leaf senescence and death - regulation by multiple layers of control and implications for aging in general. *Journal of Cell Science*, 2013 126, 4823-4833.

Yalçın, G; Vardar, F. (2016). The alleviating effects of salicylic acid application against aluminium toxicity in barley (*Hordeum vulgare* L.) roots. *Biologia*, 71, 1338-1344.

Yao, S; Huang, W; Pan, C; Zhan, J; He, LF. (2016). Caspase-like proteases regulate aluminum-induced programmed cell death in peanut. *Plant Cell Tissue and Organ Culture*, 127, 691-703.

Yu, Y; Jin, C; Sun, C; Wang, J; Ye, Y; Zhou, W; Lu, L; Lin, X. (2016). Inhibition of ethylene production by putrescine alleviates aluminium induced root inhibition in wheat plants. *Nature Scientific Reports*, 6, 18888.

Zhan, J; Li, W; He, HY; Li, CZ; He, LF. (2014). Mitochondrial alterations during Al-induced PCD in peanut root tips. *Plant Physiology and Biochemistry*, 75, 105-113.

Zhang, WH; Rengel, Z; KUO, J; Yan G. (1999). Aluminium effects on pollen germination and tube growth of *Chamelaucium uncinatum*. A comparison with other Ca^{2+} antagonists. *Annals of Botany*, 84, 559-564.

Zhou, XX; Yang, LT; Qi, YP; Guo, P; Chen, LS. (2015). Mechanisms on boron-induced alleviation of aluminum-toxicity in *Citrus grandis* seedlings at a transcriptional level revealed by cDNA AFLP analysis. *PLOS ONE*, 10, e0115485.

Zhu, CQ; Zhang, JH; Sun, LM; Zhu, LF; Abliz, B; Hu, WJ; Zhong, C; Bai1, ZG; Sajid, H; Cao, XC; Jin, QY. (2018). Hydrogen sulfide alleviates aluminum toxicity via decreasing apoplast and symplast Al contents in rice. *Frontiers in Plant Science*, 9, 294.

Zhu, XF; Zhao, XS; Wang, B; Wu, Q; Shen, RF. (2017). Elevated carbon dioxide alleviates aluminum toxicity by decreasing cell wall hemicellulose in rice (*Oryza sativa*). *Frontiers in Physiology*, 8:512.

In: Advances in Health and Natural Sciences ISBN: 978-1-53614-639-4
Editors: Burcu Yuksel et al.

Chapter 11

COMPARATIVE ANALYSIS OF DIFFERENT REGIONS OF RAT BRAIN TISSUE BY FT-IR SPECTROSCOPY

Sevgi Türker Kaya, Dilan Yıldırım, Didem Şentürk, Kübra Dağcı and Gül İlbay
Department of Biology, Faculty of Arts and Sciences,
Kocaeli University, Kocaeli/Turkey
Department of Physiology, Faculty of Medicine,
Kocaeli University, Kocaeli/Turkey

ABSTRACT

The brain is a very complex system of millions of nerve cells and consists of anatomically and functionally independent structures, all of which having many functions to control living organisms. The complexity of the brain tissue arises from its heterogeneity, distinct structural ordering, chemical composition and functional activity. Due to such complexity, region specific investigations of brain tissue may have prominent importance for identification of pathological conditions. In the present study, whole brains of Wistar rats and different brain regions such as right-left hemisphere, hippocampus and cerebellum were comparatively analyzed by Fourier transform infrared (FT-IR) spectroscopy. Within the scope of detailed spectral analysis, the frequency and areas of some specific modes such as olefinic, CH_2 asymmetric and symmetric, amide I, C=O and PO^-_2 asymmetric stretchings were evaluated. In addition, the bandwidths of CH_2 asymmetric and symmetric bands were analyzed. Also, olefinic = CH/lipid, CH_2/lipid, lipid/protein ratios were calculated. Thus, differences in parameters such as protein and lipid amounts, lipid packing and lipid fluidity were observed. As a result, corresponding findings support that FT-IR spectroscopy can be successfully applied for comparative analysis of distinct parts of the same tissue.

Keywords: brain, hemisphere, cerebellum, hippocampus, FT-IR

INTRODUCTION

The brain serves as the central organ of the nervous system; and together with the spinal cord, it constitutes the central nervous system in all vertebrates and most invertebrates. The brain controls most of the activities of the body such as processing, integration, and coordinating the information as well as receiving data from sensory organs and making decisions as to the instructions sent to the rest of the body (Krafft et al., 2006; Hackett et al., 2011). As having a very important role for most of the animals including humans, the brain tissue represents very complex structure with varied anatomic areas.

In general, the brain consists of right and left hemispheres, which results in hemispheric asymmetry defined as the presence of functional, anatomic or neurochemical differences between distinct hemispheres in a variety of species including rodents and humans (Reddon et al., 2009). (Hering-Hanit et al., 2001; Avci et al., 2008; Gökçer et al., 2012). Although two hemispheres are specialized in different functions, they are not isolated from each other, and research has focused on how they cooperate. Molecular and functional imaging techniques have revealed evidence of neurochemical asymmetries in brain and provided information about the relationship between brain lateralization and animal behavior and function (Sang et al., 2015).

A major component of the brains of humans and other vertebrates is the hippocampus located in the medial part of the temporal lobe extending along the temporal horn of the lateral ventricle (Songur et al., 2001; Spring et al., 2009; Spasojevic et al., 2013; İzci et al., 2015). The hippocampus belongs to the limbic system. Despite its close relationship with many other regions of brain makes hippocampal function difficult to define, it plays important roles in the consolidation of information from short-term memory to long-term memory, and in spatial memory that enables navigation (Yassa et al., 2011). Since different neuronal cell types are organized into layers in the hippocampus, it has frequently been used as a model system for studying neurophysiology, therefore; this region is the most studied one on compared to other brain regions.

The second major part of the brain is the cerebellum, which consists of two small half-cylinders attached to the back of the brain and at the front of the spinal cord. It is located in the posterior fossa and occupies 10% of the brain's volume (Derman et al., 1986). This specialized part is well-known for its role in coordinating motor output and adapting involuntary repetitions to support sensory pre-defined hat-based learning (Parmar et al., 2018) and in motor, cognitive and behavioral development as obtained from the evidence from the fetal, neonatal and pediatric populations (Andreottia et al., 2018).

Regional differentiation of brain tissue results from region specific structural and functional variations of biomolecules being major building blocks of brain (Krafft et al., 2006; Hackett et al., 2011). Thus, the investigation of such alterations depending on distinct parts of the brain can be pivotal for understanding of the mechanisms of different diseases and disorders. (Samuel et al., 2008; Engel et al., 2013; Dixit et al., 2016). For this purpose, infrared (IR) spectroscopy is very suitable for analysis of brain tissue due to its ability in detection of the changes on content, structure and dynamics of biomolecules in cells and tissues (Cakmak et al., 2012; Derenne et al., 2012; Nallala et al., 2013; Ozek et al., 2014; Turker-Kaya et al., 2016). IR spectroscopy does not require any staining or complicated sample preparation (Derenne et al., 2012; Turker et al., 2014a). An IR spectrum of a biological sample is the sum of various contributions from lipids, proteins, carbohydrates, nucleic acids and all other chemical species (Carmona et al., 2008; Miller et al., 2013; Ozek et al., 2014; Turker et al., 2014a,b). The intensities and/or areas provide quantitative information, peak positions relate to physical states of molecules, and bandwidth gives dynamic information (Severcan et al., 2005; Carmona et al., 2008; Miller et al., 2013; Ozek et al., 2014; Kumar et al., 2014; Turker et al., 2014a;b; Turker-Kaya et al., 2016). Due to such advantages, FT-IR spectroscopy has been increasingly employed to investigate molecular variations between different tissues and alterations induced by drugs, and even disease states which are not easily detectable by morphological methods in cells and tissues (Berger et al., 2010; Leskovjan et al., 2010; Cakmak et al., 2011; Bozkurt et al., 2012; Travo et al., 2012; Bellisola et al., 2012; Turker et al., 2014a;b; Turker-Kaya et al., 2016).

In the present study, we aimed to analyze left and right hemisphere, hippocampus and cerebellum regions of rat brain tissue in comparison to whole brain by FT-IR spectroscopy as similarly performed in our previous studies (Turker et al., 2014a;b; Turker-Kaya et al., 2016; Turker-Kaya et al., 2017). We monitored the relative changes in lipid and proteins content, membrane lipid packing, membrane fluidity, and lipid peroxidation products (LPPs), all of which are very important parameters for proper functioning of brain.

Material and Methods

A. Animal Experiments

Adult Wistar rats were used for animal experiments carried out at Kocaeli University/Turkey. The animals were kept in a constantly warm room for 12 hours in the light-12 hour dark cycle with 60% relative humidity. The rats were decapitated and their brains removed out under deep ether anesthesia. The dissected brains were separated into 4 groups as whole brain (WB, n = 7), left hemisphere (LH, n = 6), right hemisphere (RH, n = 6), hippocampus (HP, n = 6) and cerebellum (CR, n = 6). The removed brain tissues were stored at - 80°C deep freeze till spectroscopic analysis. All procedures involving animals were in compliance with the European Community Council Directive of 24 November 1986, and ethical approval was granted by the Kocaeli University Ethics Committee (Number: 2017/091 and 2018/040HD, Kocaeli, Turkey).

B. FT-IR Spectroscopy Studies

For sample preparation, earlier studies were followed (Elibol-Can et al., 2011; Turker-Kaya et al., 2016). Briefly, freeze-drying was performed using a (Labconco FreeZone®, Benchtop Freeze Dry System Model 77520) for 24 hours to remove excess water from brain tissues. Fully dried tissues were ground and each tissue powder (2 mg) was mixed with 100 mg of pre-dried Potassium Bromide (KBr). Later, the mixture was dried again in the freeze drier for 18 hours to remove all traces of remaining water. The prepared mixture was applied placed a pressure of 100 kg/cm^2 for 5 minutes to obtain a solid KBr pellet.

A Perkin-Elmer Spectrum One FT-IR spectrometer (Perkin-Elmer Inc., Norwalk, CT, USA) equipped with a MIR TGS detector was utilized for spectral acquisition. All spectra were recorded in the 4000–450 cm^{-1} region at room temperature. Each interferogram was collected with 100 scans at 4 cm^{-1} resolution.

Although special effort was given to produce same thickness of each KBr pellet, to minimize intra-sample variability which may be due to variations in experimental conditions, two independent pellets prepared by taking separate portions randomly from the same sample were scanned. The average spectrum of these two replicates giving identical spectra representing one animal was used in detailed data analysis.

C. Data Analysis

Each spectrum was analyzed by the Perkin Elmer Spectrum One program to reveal the spectral differences between the groups in terms of band area, band position and band width. The band positions (wavenumber) values were calculated by taking 70% of the peak height by evaluating the center of gravity of the bands. The bandwidth was calculated by converting peak height at 70% to cm^{-1}. The peak area was calculated as related to a linear baseline between two baseline points which involve the maximum peak height. The baseline point was identified according to start and end points of the peak. Later, the corrected area which is the area between the spectrum and the marked baseline within the marker bar limits was found the peak area value of the peak of interest.

Taking possible interference due to detectable variability of sample thickness into consideration, area ratios of some specific IR modes were also evaluated to approximate content changes in biomolecules. The lipid/protein ratio was obtained by proportioning the total area under the CH_2 asymmetric and the CH_2 symmetric stretching (2925 cm^{-1} and 2852 cm^{-1}) bands to the total area under the bands amide I and amide II (1645 cm^{-1} and 1547 cm^{-1}). For olefinic = CH/lipid and CH_2/lipid ratios, the areas under the olefinic = CH (3012 cm^{-1}) and the CH_2 asymmetric stretching (2925 cm^{-1}) modes were divided to the total saturated lipid (the sum of the area under the CH_2 asymmetric and symmetric stretching bands).

RESULT AND DISCUSSION

As being major components of brain tissue, any change in the structure and function of biomolecules such as lipids and proteins directly affects the proper functioning of the neural network.

In the current study, we analyzed distinct regions of brain, each of which contains various cellular and molecular components. In this complexity FT-IR spectroscopy is insufficient to provide information about specific components. However, as a result of the Conne advantage, the vibrations of functional groups in different biomolecules appear at distinct wavenumber values in an FT-IR spectrum. This enables separation of the peaks from different biomolecules to be analyzed even in weak absorption modes (Gasper et al. 2009; Leskovjan et al. 2010; Elibol-Can et al. 2011; Turker et al. 2014b). Therefore, it is possible to monitor saturated lipids, unsaturated lipids, cholesterol ester, triglycerides, proteins, nucleic acids and carbohydrates. The present study was conducted to comparatively analyze whole brain, right and left hemispheres, hippocampus and cerebellum sections of Wistar rats. To achieve this, we monitored the variations in the frequencies, bandwidths and peak areas of some vibrational modes of biomolecules as previously performed in other reports (Cakmak et al. 2011; Ozek et al. 2014; Turker-Kaya et al. 2016).

According to Beer–Lambert law, the intensity and area of an absorption band in an FT-IR spectrum offers relative information on the content of corresponding functional groups within complex tissue systems (Berger et al. 2010; Leskovjan et al. 2010). This has been previously confirmed by biochemical assays for LPPs (Severcan et al. 2005; Leskovjan et al. 2012; Turker et al. 2014a), proteins (Bozkurt et al. 2012) and lipids (Derenne et al. 2012). By utilizing this, in the current study, we evaluated area ratios of some modes to predict the changes in the content of these molecules without need of biochemical assays.

An FT-IR spectrum obtained from rat brain tissue sample is shown in Figure 1. As demonstrated in the figure, there are many bands of at different wavenumbers in the spectrum of the brain tissue, so the spectrum is quite complex. Table 1 illustrates assignment of each infrared band based on the literature (Wetzel et al., 1999; Toyran et al., 2004; Demir et al., 2015).

In order to show the spectral differences among the groups in more detail, spectra were divided into two separate regions, 3050-2800 cm^{-1} and 1800-900 cm^{-1}. Figure 2 and 3 illustrate average spectra of WB, LB, RB, CR, and HP groups in 3050-2800 cm^{-1} and 1800-900 cm^{-1} region, respectively. Table 2 represents numerical results of spectral analysis of each group.

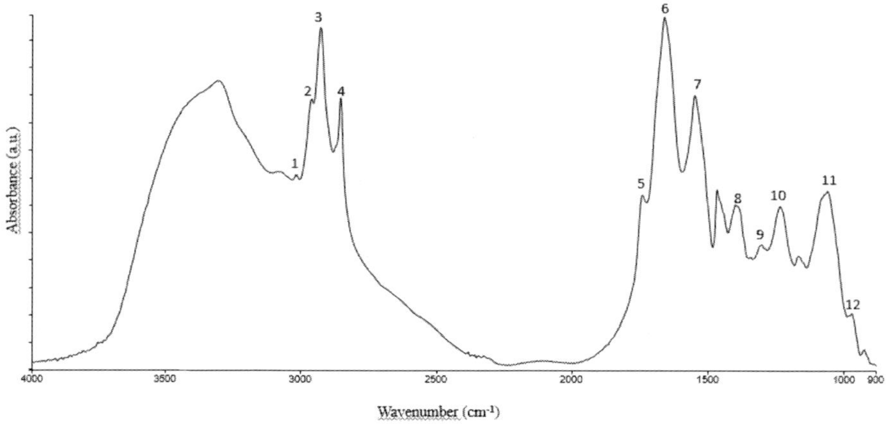

Figure 1. An average FT-IR spectrum obtained from rat brain tissue sample in 4000-900 cm^{-1} wavenumber range.

Table 1. General band assignments of average FT-IR spectrum of rat brain based on the literature

No	Frequency (cm^{-1})	Definition of the spectral assignments
1	3012	Olefinic = CH stretching: unsaturated lipid
2	2956	CH_3 asymmetric stretching: mainly lipid with little contribution from protein, carbohydrate and nucleic acid
3	2925	CH_2 asymmetric stretching: mainly lipids with little contribution from protein, carbohydrate and nucleic acid
4	2852	CH_2 symmetric stretching: mainly lipid with the little contribution from protein, carbohydrate and nucleic acid
5	1740	Saturated ester C=O stretch: phospholipid, cholesterol esters triglyceride
6	1645	Amide I (protein C=O stretch): protein
7	1547	Amide II (C=N and N–H stretching): protein
8	1400	COO^- symmetric stretching: fatty acid
9	1306	Amide III: protein
10	1236	PO_2^- asymmetric stretching: phospholipid and nucleic acid
11	1080	PO_2^- symmetric stretching: phospholipid and nucleic acid
12	970	C^+–N–C stretching: nucleic acid, ribose phosphate main chain vibrations of RNA

The myelin sheath, a phospholipid layer surrounding the axons of neurons, has a very high lipid content due to its multilayer structure and a greater lipid-protein ratio than most cellular compartments. Thus, the lipid bands observed in the brain tissue may originate from the myelin sheaths. For this reason, it is important to define such differences in lipid molecules that may cause deterioration in signal transmission in the brain (Palus et al., 2017). Similarly, brain tissue depends on oxidative pathway to utilize fuels, and has high content of unsaturated lipids when compared with other organs. This also brings about high level of mitochondria where reactive oxygen species are produced (Adam-Vizi et al., 2005). These molecules readily attack lipids and lipid peroxidation occurs, and LPPs are released into extra and intracellular site of cell (Cakmak et al. 2011; Turker et al. 2014a). Therefore, the detection of LPPs may provide information about not only possible pathogenesis of any condition but also antioxidant capacity of an agent in tissues, also shown by earlier studies (Battisti et al. 2008; Ramana et al. 2014). For this reason, in the current study, we specifically compared brain regions in terms of content of unsaturated lipids and presence of LEPs. In Figure 1, the olefinic band at 3012 cm^{-1} originates from the C-H stretching mode of the HC = CH groups. It gives information about the degree of unsaturation of the acyl chains, particularly in the membrane lipids and lipid peroxidation (Turker-Kaya et al., 2016). Normally, when a decrease in unsaturated lipids occurs, a lower value in both olefinic mode area and olefinic/lipid ratio is expected. However, when reduced amount of unsaturated lipids is caused by lipid peroxidation, presence of LEPs results in an increase in olefinic mode area and olefinic/lipid ratio is observed as reported earlier (Turker et al. 2014a). This may be due to the loss of unsaturation being compensated by the presence of = CH groups in LEPs such as lipid aldehydes and alkyl radicals (Liu et al., 2002; Turker-Kaya et al., 2016). The ratio values of olefinic/lipid for all groups are given in Table 2. As seen from the table, when compared to the others, the lowest value was obtained in CR. The obtained finding may be resulted from low level of unsaturated and LEPs and high content of saturated lipids and/or both. On the other hand, while HP was same with WB, there was no difference between RH and LH. This shows that the unsaturated lipids and LEPs are accumulated least in CR in contrast to other regions of brain.

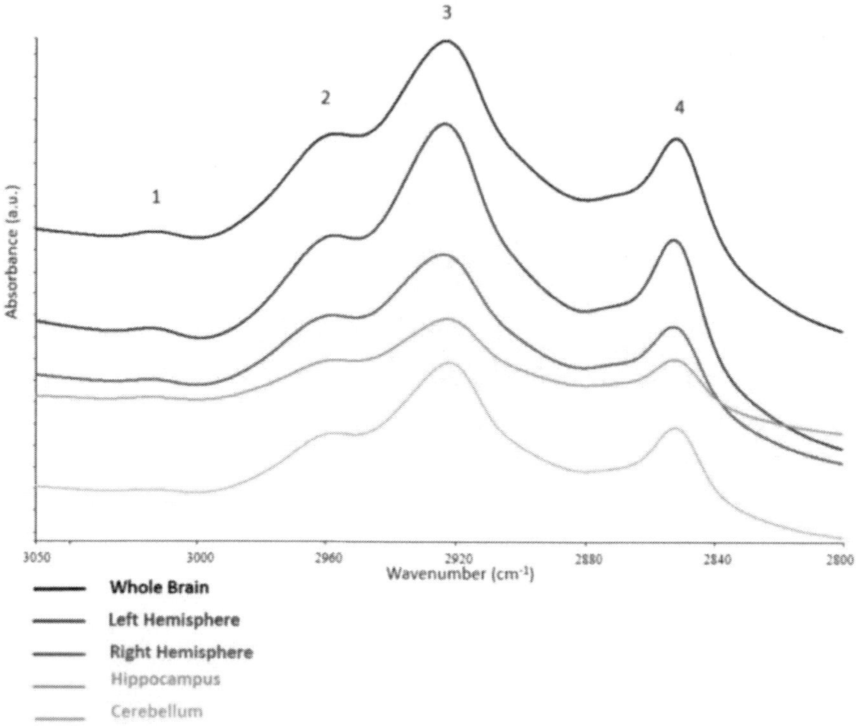

Figure 2. Representative FT-IR spectra of WB, RH, LF, CR, HP groups in the region between 3050–2800 cm^{-1}.

Membrane proteins such as membrane-bound enzymes, ion channels and receptors are dominantly present and function in nerve cell activity (Turker et al., 2014a). These types of proteins are classified as integral membrane proteins and peripheral membrane proteins. Integral membrane proteins are in contact with with fatty acyl chains of membrane lipids whereas peripheral membrane proteins are predominantly bound to lipid head groups (Boggs et al., 1987; Brasseur et al., 1997; Turker et al., 2014b). Due to these bilateral interactions between membrane lipids and proteins, physical properties of lipids and proteins mutually affect their activity (Sandermann et al., 1978; McElhaney et al., 1982; Yeagle et al., 1992). To evaluate distinct brain regions in terms of physical state of membrane lipids, which determine their packing within bilayer, we monitored the frequency changes of the CH_2 asymmetric (2925 cm^{-1}), the

CH_2 symmetric (2852 cm^{-1}), the C=O (1740 cm^{-1}), the PO_2^- asymmetric (1236 cm^{-1}) stretching bands. The CH_2 asymmetric and CH_2 symmetric stretching bands are mainly due to acyl chains of membrane lipids (saturated lipids) (Turker et al., 2014a; Turker et al., 2014b). And, the monitoring of frequency shifting of these bands give information about order and disorder state of membrane (Turker et al., 2014b). Changes in the frequency values of such modes is related with the alterations in amount of gauche or trans conformers in the acyl chains. This may further cause the membrane to be loose or tightly packed (Senak et al., 1993, Akkas et al., 2007; Turker et al., 2014a). Evaluation of the frequency values of the CH_2 asymmetric and CH_2 symmetric band revealed no significant differences among all groups (Table 2). This may mean that there is no prominent variation in membrane order and acyl chain packing parameters of the regions in compared to WB. On the other hand, in Figure 3, the C=O stretching band at 1739 cm^{-1} is a band that attributes to triglycerides, cholesterol, and carbonyl groups of phosholipids (Severcan et al., 2005; Turker et al., 2014a). The PO_2- asymmetric stretching band at 1229 cm^{-1} is bound to the phosphodiester backbone of phospholipids and cellular nucleic acids (Cakmak et al., 2006). The shifts in the C=O and the $PO2^-$ asymmetric modes to lower wavenumbers suggest an increase in the hydration state of the glycerol backbone near the hydrophilic part and polar head group of the membrane lipids. The hydrogen bonding might be between water molecules and the oxygen molecules of both carbonyl and phosphate groups of phospholipids or hydrophilic protein residues (Akkas et al., 2007). As seen in Table 2, the C=O frequency value shifted to higher values in both RH and LH but lower values HP and CR when compared to WB. On the other hand, any significant shifting in PO_2- asymmetric frequency value was not observed among the groups. These corresponding findings may mean that physical state of glycerol backbone near the hydrophilic part of membrane lipids show variations depending on specific brain regions. Nevertheless, state of acyl chain and phosphate groups of membrane lipids do not differ in distinct part of investigated regions of brain. Such parameter also reflects packing parameter of acyl chains of membranes.

Table 2. The values are the mean ± standard deviation for each sample

	WB (n = 7)	RH (n = 6)	LH (n = 6)	HP (n = 6)	CR (n = 6)
Frequency					
CH_2 asymmetric	2924.42 ± 0.05	2923.88 ± 0.07	2924.49 ± 0.08	2923.32 ± 0.07	2923.22 ± 0.13
CH_2 symmetric	2852.48 ± 0.05	2852.17 ± 0.02	2852.06 ± 0.03	2852.27 ± 0.05	2852.45 ± 0.03
Amide I	1655.29 ± 0.09	1652.71 ± 0.22	1652.53 ± 0.14	1653.04 ± 0.12	1652.71 ± 0.3
C=O	1739.59 ± 0.02	1745.65 ± 0.09	1745.8 ± 0.07	1737.52 ± 0.05	1736.58 ± 0.07
PO^-_2	1235.33 ± 0.19	1233.74 ± 0.12	1234.7 ± 0.12	1235.32 ± 0.17	1234.28 ± 0.17
Band area ratio					
Olefinic/lipid	0.08 ± 0.004	0.10 ± 0.02	0.10 ± 0.003	0.08 ± 0.004	0.07 ± 0.012
CH_2 asym/lipid	0.33 ± 0.012	0.36 ±0.019	0.38 ± 0.020	0.36 ± 0.04	0.39 ± 0.05
Lipid/protein	0.99 ± 0.07	0.85 ±0.014	1.26 ± 0.07	1.01 ± 0.09	0.73 ± 0.09
Bandwidth					
CH_2 asymmetric	22.85 ± 0.16	20.67 ± 0.14	21.47 ± 0.11	21.41 ± 0.14	20.52 ± 0.04
CH_2 symmetric	9.82 ± 0.06	8.14 ± 0.03	8.14 ± 0.04	9.71 ± 0.005	9.71 ± 0.008

Changes in the fluidity of the cell membrane can be determined by investigating the bandwidth values of the CH_2 asymmetric stretching mode (Severcan et al., 2010, Turker-Kaya et al., 2016). The bandwidth value of this band gives information about the dynamics of the system (Turker et al., 2014b). As illustrated in Table 2, the highest value of CH_2 asymmetric bandwidth values is observed in WB. The obtained lower values in other regions may imply reduced lipid fluidity (Turker-Kaya et al., 2016). Here, it should be emphasized that this result related with fluidity is about all membraneous structures in brain rather than specific subcellular membranes of organelles. In other words, our samples contain all membrane compartments in addition to cell membrane. For that reason, the obtained outcome reflects cumulative fluidity parameter. Decrease in fluidity might have resulted from variations in unsaturated lipids, which in turn causing the formation of kinks and force the molecules further apart and allow for more movement. Thus, a lowered membrane fluidity found in the current study can be another indication of functional changes of membrane proteins.

CH_2/lipid ratio was calculated to determine the alterations in the membrane phospholipids chain length (Cakmak et al., 2016). When the ratio of CH_2/lipid is examined in Table 2, it is seen that this value is similar in the HP and RH, while the highest value is in the CR and the lowest value is in the WB.

The bands at 1645 cm^{-1} and 1540 cm^{-1} in Figure 3 correspond to amide I and amide II bands of proteins, respectively. Amide I is mainly caused by C-O stretching of proteins while amide II is rassigned to N-H and C-N stretching (Cakmak et al., 2006; Turker et al., 2014a). They are used to determine protein structure or conformation and functional differences (Yan et al., 2014). When the 1650 cm^{-1} amide I band frequency values are examined in Table 2, the highest value was observed in WB, then the HP value came. The RH and CR tissues were similar. The lowest frequency value was observed in the LH. This means that there are structural variations in protein among brain regions.

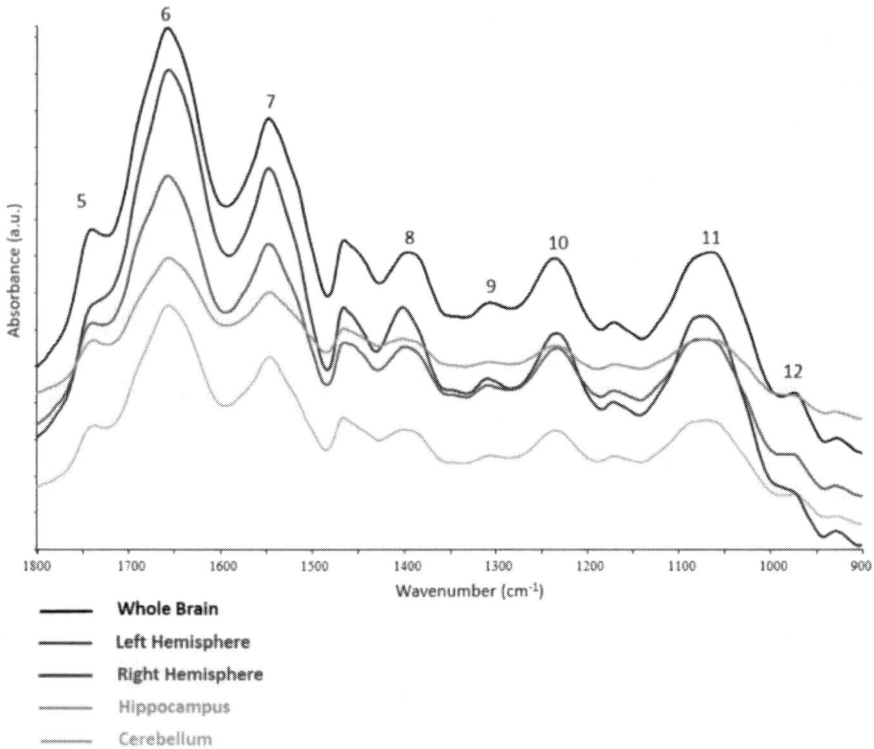

Figure 3. Representative FT-IR spectra of WB, RH, LF, C, H groups in the region between 1800–900 cm^{-1}.

When membrane lipids and proteins are thought to be in contact with each other, any change in the ratio of these molecules can alter membrane thickness and affect lipid regulation and fluidity (Awayda et al., 2004; Cakmak et al., 2011; Turker et al., 2014a). In this context, we calculated the lipid/protein ratio. As shown in Table 2, this ratio is similar in WB and HP tissues and the highest value is in the LH tissue. The lowest values belong to the right RH and CR tissues.

As a result, when all these differences are taken into account, changes in fatty acids, protein conformation, and ratios in different brain tissues support neurochemical asymmetry studies between brain regions. At the same time, the parameters of the right-left hemisphere, cerebellum, and hippocampus tissues may be related to the functions of the brain parts, which are different from whole brain tissue parameters. However, the relationship between neurochemical changes and functions is not yet known.

REFERENCES

[1] Eric, MR; Kewei, C; Daniel, B. Functional brain abnormalities in young adults at genetic risk for late-onset Alzheimer's dementia, *Proceeding of the National Academy of Sciences of the United States of America*, 2003, 101, 284-289.

[2] Mink, JW; Blumenschine, RJ; Adams, DB. Ratio of central nervous system to body metabolism in vertebrates: its constancy and functional basis, *American Journal of Physiology-Regulatory*, 1981, 241, R203-R212.

[3] Krafft, C; Sergo. Biomedical applications of Raman and infrared spectroscopy to diagnose tissues, *Vibrational Spectroscopy*, 2006, 20, 195-218.

[4] Hackett, MJ; McQuillan, JA; El-Assaad, F; Aitken, JB; Levina, A; Cohen, DD; Siegele, R; Carter, EA; Grau, GE; Hunt, NH; Lay, PA. *Analyst*, 2011, 136, 2941.

[5] McIlwain, H; Bachelard, HS. *Biochemistry and the Central Nervous System*, Edinburgh: Churchill Livingstone, *Biochemical Education*, 1985, 14.

[6] Boggs, JM. Lipid molecular hydrogen bonding: influence on structural organization and membrane function, *Biochimica et Biophysica Acta*, 1987, 906, 353–404.

[7] Kubota, T; Fujisaki, K; Itoh, Y; Yano, T; Sendo, T; Oishi, R; Apoptotic injury in cultured human hepatocytes induced by HMG-CoA reductase inhibitors, *Biochemical Pharmacology*, 2004, 67, 2175–2186.

[8] Turker, S; İlbay, G; Severcan, M; Severcan, F. Investigation of Compositional, Structural, and Dynamical Changes of

Pentylenetetrazol-Induced Seizures on a Rat Brain by FT-IR Spectroscopy, *Analytical Chemistry*, 2014, 86, 1395−1403.

[9] Cakmak, G; Togan, I; Severcan, F. 17Beta-estradiol induced compositional, structural and functional changes in rainbow trout liver, revealed by FT-IR spectroscopy: a comparative study with nonylphenol, *Aquatic Toxicology*, 2006, 77, 53–63.

[10] Turker-Kaya, S; Mutlu, O; Çelikyurt, İ; Akar, F; Ulak, G. Tianeptine, olanzapine and fluoxetine show similar restoring effects on stress induced molecular changes in mice brain: An FT-IR study, *Spectrochimica Acta Part A: Molecular and Biomolecular Spectroscopy*, 2016, 161, 178–185.

[11] Severcan, F; Bozkurt, O; Gurbanov, R; Gorgulu, G. FT-IR-Spectroscopy in Diagnosis of Diabetes in Rat Animal Mode, *Journal of Biophotonics*, 2010, 3, 621–631.

[12] Yusuf, İ; Yahya Cem, E. *Hippocampus: Its Structure and Functions, Türk Nöroşir Dergisi*, 2015, 25, 287-295.

[13] Ahmet, S; Oğuz Aslan, Ö; Mustafa, S. *Hippocampus, Türkiye Klinikleri Tıp Bilimleri Dergisi*, 2001, 21, 427-431.

[14] Vasiliki, K; Catherine, RD. Transient vimentin expression during the embryonic development of the chicken cerebellum, *International Journal of Developmental Neuroscience*, 2018, 65, 11–20.

[15] Julia, PA; Pedro, HDMP; Luiz, AVM; Marco, AR; Akiva, M; Alexander, Birbraira. Neurogenesis in the postnatal cerebellum after injury, *International Journal of Developmental Neuroscience*, 2018, 67, 33–36.

[16] Catherine, JS; Catherine, L. Structure function relationships in the developing cerebellum: Evidence from early-life cerebellar injury and neurodevelopmental disorders, *Seminars in Fetal & Neonatal Medicine*, 2016, 21, 356-364.

[17] Marcus, D'S; Jessica, B; Christiane, W; Jaume, Sastre-Garrigag. The role of the cerebellum in multiple sclerosis—150 years after Charcot, *Neuroscience and Biobehavioral Reviews*, 2018, 89, 85–98.

[18] Mingmei, B; Guixin, Q; Zewei, Sun; Guohui, Long. Relationship between Molecular Structure Characteristics of Feed Proteins and

Protein *In vitro* Digestibility and Solubility, *Asian-Australasian Journal of Animal Sciences*, 2016, 29, 1159-1165.

[19] Velmurugan, B; Senthilkumaar, P; Karthikeyan, S. Toxicity impact of fenvalerate on the gill tissue of Oreochromis mossambicus with respect to biochemical changes utilizing FTIR and principal component analysis, *Journal of Biological Physics*, 2018, 1007, 9484-9.

[20] Régis, G; Janique, D; Robert, Kissc; Tatjan, Mijatovicb; Erik, Goormaghtigh. IR spectroscopy as a new tool for evidencing antitumor drug signatures, *Biochimica et Biophysica Acta - Biomembranes*, 2009, 1788, 1263-1270.

[21] Yan, X; Khan, NA; Zhang, F; Yang, L; Yu, P. Microwave irradiation induced changes in protein molecular structures of barley grains: relationship to changes in protein chemical profile, protein subfractions, and digestion in dairy cows. *J. Agric. Food Chem*, 2014, 62, 6546-6555.

[22] Peng, QH; Khan, NA; Wang, Z; Yu, PQ. Relationship of feeds protein structural makeup in common Prairie feeds with protein solubility, *in situ* ruminal degradation and intestinal digestibility, *Animal Feed Science and Technology*, 2014, 194, 58-70.

[23] Manuel, R; Isabel, P; Francisco, V; Marc, G; Francisco, A. Neuropeptides, Neuropeptidases and Brain Asymmetry. *Current Protein & Peptide Science*, 2004, 5, 497 – 506.

[24] Spasojevic, N; Jovanovic, P; Dronjak, S; Molecular basis of chronic stress-induced hippocampal lateral asymmetry in rats and impact on learning and memory, *Acta Physiologica Hungarica*, 2013, 100, 388–394.

[25] Spring, S; Lerch, J; Wetzel, M; Evans, A; Henkelman, R. Cerebral asymmetries in 12-week-old C57Bl/6J mice measured by magnetic resonance imaging. *NeuroImage*, 2010, 50, 409–415.

[26] Samuel, CW; Zuzana, H; Lİeuko, N; Gary, PD. Asymmetrical distributions of muscarinic receptor binding in the hippocampus of female rats. *European Journal of Pharmacology*, 2008, 588, 248–250.

[27] Sang, C; Eun, JY; Sang, EK. Asymmetry of Dopamine D2/3 Receptor Availability in Dorsal Putamen and Body Mass Index in Non-obese Healthy Males. *Experimental Neurobiology*, 2015, 24, 90-94.

[28] Reddon, A; Gutiérrez-Ibáñez, C; Wylie, DR; Hurd, P. The relationship between growth, brain asymmetry and behavioural lateralization in a cichlid fish, *Behavioural Brain Research*, 2009, 201, 223-228.

[29] Hering-Hanit, R; Achiron, R; Lipitz, S; Achiron, A. Asymmetry of fetal cerebral hemispheres: in utero ultrasound study. *Archives of Disease in Childhood Fetal and Neonatal Edition*, 2001, 85, 194-196.

[30] Gökçer, E; İlker, Y; Ümmühan, A. El Tercihinin Görsel P300 Yanıtları ve Görsel Tarama Yolları Üzerine Etkisi, İnönü Üniversitesi Sağlık Bilimleri Dergisi, 2012, 1, 32-40.

[31] Avcı, DE; Yağbasan, R. *Gazi Eğitim Fakültesi Dergisi*, [*Gazi Education Faculty Journal*] 2008, 1-17.

[32] Severcan, F; Haris, PI. *Applications of Vibrational Spectroscopy in Diagnosis and Screening*, IOS Press, 2012, 100, 105-9.

[33] Cakmak, G; Zorlu, F; Severcan, M; Servercan, F. Screening of protective effect of amifostine on radiation-induced structural and functional variations in rat liver microsomal membranes by FT-IR spectroscopy, *Analytical Chemistry*, 2011, 7, 2438-2944.

[34] Ozek, N; Sara, Y; Onur, R; Severcan, F. Low dose simvastatin induces compositional, structural and dynamic changes in rat skeletal extensor digitorum longus muscle tissue, *Bioscience Reports*, 2010, 30, 41-50.

[35] Toyran, N; Zorlu, F; Dönmez, G; Severcan, F. Chronic hypoperfusion alters the content and structure of proteins and lipids of rat brain homogenates: a Fourier transform infrared spectroscopy study, *European Biophysics Journal*, 2004, 33, 549-554.

[36] Wetzel, DL; Lah, SM. Imaging Molecular Chemistry with Infrared Microscopy, *Microspectroscopy*, 1999, 285, 1224-1225.

[37] Demir, P; Akkaş, SB; Severcan, M; Zorlu, F; Severcan, F. Ionizing Radiation Induces Structural and Functional Damage on the Molecules of Rat Brain Homogenate Membranes: A Fourier Transform Infrared (FT-IR), *Spectroscopic Study*, 2015, 69, 154-64.

[38] Türker, S; Severcan, M; İlbay, G; Severcan, F. Epileptic seizures induce structural and functional alterations on brain tissue membranes, *Biochimica et Biophysica Acta*, 2014, 1838, 3088–3096.

[39] Palus, JZ; Wesełucha-Birczynska, A; Sacharz, J; Lewandowski, MH; Palus, K; Chrobok, L; Kowalski, R; Moskal, L; Birczynska, M; Sozanska, A. 2D correlation Raman microspectroscopy of chosen parts of rat's brain tissue, *Journal of Molecular Structure*, 2017, 1147, 310-316.

[40] Liu, K; Boseb, R; Mantsch, H. Infrared spectroscopic study of diabetic platelets, *Vibrational Spectroscopy*, 2002, 28, 131–136.

[41] Cakmak, G; Zorlu, F; Severcan, M; Severcan, F. Screening of protective effect of amifostine on radiation-induced structural and functional variations in rat liver microsomal membranes by FT-IR spectroscopy, *Analytical Chemistry*, 2011, 83, 2438-2444.

[42] Akkas, SB; Inci, S; Zorlu, F; Severcan, F. Melatonin affects the order, dynamics and hydration of brain membrane lipids, *Journal of Molecular Structure*, 2007, 834–836, 207-215.

[43] Senak, L; Mendelsohn, R. CH_2 wagging modes as quantitative ır probes of acyl chain conformational order in phospholipid membranes, *Biophysical Journal*, 1993, 121, A299-A299.

[44] Saberwal, G; Nagaraj, R. Cell-lytic and antibacterial peptides that act by perturbing the barrier function of membranes: facets of their conformational features, structurefunction correlations and membrane-perturbing abilities, *Biochimica et Biophysica Acta*, 1994, 1197, 109–131.

[45] Gasper, R; Dewelle, J; Kiss, R; Mijatovic, T. IR spectroscopy as a new tool for evidencing antitumor drug signatures, *Biochimica et Biophysica Acta*, 2009, 1788, 1263–1270.

[46] Leskovjan, A; Kretlow, A; Miller, L. Fourier transform infrared imaging shows reduced unsaturated lipid content in the hippocampus

of a mouse model of Alzheimer's disease, *Analytical Chemistry*, 2010, 82, 2711–2716.

[47] Elibol-Can, B; Jakubowska-Dogru, E; Severcan, M; Severcan, F. The effects of short-term chronic ethanol intoxication and ethanol withdrawal on the molecular composition of the rat hippocampus by FT-IR spectroscopy, *Alcoholism: Clinical and Experimental Research*, 2011, 35, 2050–2062.

[48] Awayda, MS; Shao, W; Guo, F; Zeidel, M; Hill, WG. ENaC–membrane ınteractions regulation of channel activity by membrane order, *The Journal of General Physiology*, 2004, 123, 709-727.

[49] Cakmak, G; Severcan, M; Zorlu, F; Severcan, F. Structural and functional damages of whole body ionizing radiation on rat brain homogenate membranes and protective effect of amifostine, *International Journal of Radiation Biology*, 2016, 92, 837-848.

[50] Adam-Vizi, V. Production of reactive oxygen species in brain mitochondria: contribution by electron transport chain and non-electron transport chain sources, *Antioxid Redox Signal*, 2005, 7, 1140-9.

[51] Boggs, JM. Lipid molecular hydrogen bonding: influence on structural organization and membrane function, *Biochimica et Biophysica Acta*, 1987, 906, 353–404.

[52] Brasseur, G; Pillot, T; Lins, L; Vandekerckhove, JM. Peptides x inmembranes: tipping the balance of membrane stability, *Trends in Biochemical Sciences*, 1997, 22, 167, 171.

[53] Yeagle, P. *The Structure of Biological Membranes*, CRC Press, Boca Raton, FL, 1992.

[54] Sandermann, H. *Biochimica et Biophysica Acta*, 1978, 515, 209–237.

[55] McElhaney, RN; Razin, S; Rottem, S. *Current Topics in Membranes and Transport*, Academic Press, 1982, 17.

In: Advances in Health and Natural Sciences ISBN: 978-1-53614-639-4
Editors: Burcu Yuksel et al.

Chapter 12

SINGLE CELL GEL ELECTROPHORESIS FOR DETERMINATION OF PLANT RESPONSES TO GENOTOXIC STRESS

Yıldız Aydın[1,*], Ahu Altınkut Uncuoğlu[2] and Filiz Vardar[1]

[1]Department of Biology, Faculty of Science and Letters, Marmara University, Istanbul, Turkey

[2]Department of Bioengineering, Faculty of Engineering, Marmara University, Istanbul, Turkey

ABSTRACT

The transmission of the energy to the higher levels is through the food chains, starting from the main source plants. There is a balance between all the livings in the ecosystem in which the plants are the first ring of this chain. Environmental mutagens and other related substances threaten humans and all other living organisms in an ecosystem. Plants are exposed to various types of environmental agents such as agricultural pesticides and plant growth regulators, or accidentally present in polluted

* Corresponding Author Email: ayildiz@marmara.edu.tr.

air, soil or water. While pesticides used against harmful effects of agricultural production affect the target organisms, they can affect the non-target organisms through wrong and especially excessive applications. The adverse effects of toxic agents that accumulate in aquatic and terrestrial crops in an ecosystem and accumulate there as a bioindicator in plants should be identified?. Plant biotests are very important methods for evaluating the genotoxic contamination in the environment. For quick determination and measurement of the genotoxicity for these stress factors various molecular and biochemical tests are required. Single cell gel electrophoresis is also known as Comet assay that is used to investigate the cell response to several biotic and abiotic genotoxic agents leading to oxidative DNA damage. The Comet assay is used as a useful tool in assessing the potential of plants as source of information on the genotoxic impact of dangerous pollutants and sensitive sensors in ecosystems. This technique is based on the quantification of denatured DNA fragments migrating out of the cell nucleus during electrophoresis. Current review aims to give information about Comet assay and its applications as a valuable tool for monitoring DNA damage in different plant systems. The single cell gel electrophoresis studies were performed by our group to determine the genotoxic effects of different concentrations of pesticides, heavy metals and different plant growth regulators used in tissue culture applications in tobacco as model organisms, cereals (barley and wheat) and sunflower.

Keywords: COMET, biomonitoring, genotoxic effect, plant

INTRODUCTION

As it is known, plants are the main source in energy transmitting to higher levels in the food chain. The first ring of this chain are plants, the next rings are herbivores and carnivorous containing animals and humans. There is a balance between all living things in an ecosystem. The change that occurs at any level causes fluctuations in the different stages, causing the distortion of the existing balance. Today, the balance in different ecosystems is affected negatively by genotoxic agents in particular. Environmental mutagens and other related substances threaten humans and all other living things in an ecosystem. For example, pesticides are one of the chemicals that all living beings in the nature are exposed to. Pesticides are harmful to other living organisms as well as the target they are applied to, but they are the most effective fighting methods used to prevent the waste of agricultural products. Therefore, with proper use, the effects on other living things outside the target organisms should be tried to be minimized.

The adverse effects of toxic agents that accumulate in end-of-year aquatic and terrestrial ecosystems and accumulate in fisheries and plants that are studied as bioindicators should be identified. It is also a necessity for human health and well-being.

Genotoxic damage has negative effects on the important physiological conditions required for healthy life such as rejuvenation of living things, genetic characteristics and population dynamics. Various methods are currently being used to determine whether the DNA molecule is at risk for mutagenic effects and genotoxicity. One of the methods used to examine the genotoxic risk ratios is the Comet test, where direct DNA damage is measured. Single cell gel electrophoresis (SCGE) or so-called "Comet assay"; is a more sensitive and reliable method that results in shorter time than conventional cytogenetic methods used in measuring genotoxic risk ratios (Gichner et al., 2003; Lin et al., 2007). The Comet method has also the advantages such as allowing direct measurement at the DNA level and detecting genetic damage in a short time and directly. Use of this technique also extends the utility of plants in basic and applied studies in environmental mutagenesis.

The DNA breaks which migrate out of the nuclei embedded in a thin agarose layer can be measured by comet assay (Menke et al., 2001). The preparation of uniform and stationary gels in the slides is crucial to ensure that the cells embedded in the gel can be retained intact until the end of the experiment phase and to remove the factors that could affect the background appearance of the samples and in order to increase the gel fixation, roped lamellae are used (Tatli et al., 2008). The slides are placed in the electrophoresis tank containing the buffer solution and allowed to strand for a while in order to unwinding the DNA. If the unwinding and electrophoresis steps are carried out in an alkaline pH environment, the breaks of the double and single strand in the DNA can be determined. If both the unwinding and the electrophoresis are more than carried out in pH 13, the base damage in the crosslinks as well as the breaks in the double and single strand can be determined. It is very important to keep the temperature constant during electrophoresis in order to avoid the differences that may occur between samples. The slides are washed at room temperature with 0,4 M Tris-HCl (pH 7,5) to neutralize the resulting alkali medium following electrophoresis and to provide a clear image under the microscope (Gichner et al., 2008b; Rank et al., 2009).

Detection of DNA single strand breaks in cells was first described by Rydberg and Johanson (1978). Cells were embedded with agarose on a microscope slide and cell lysed under lightly alkaline conditions to allow partial dissolution of DNA. The cells were stained with acridine orange after neutralization and photometrically analyzed. The proportion of double-stranded DNA to single-stranded DNA, the proportion of green fluorescence to red fluorescence, and the extent of DNA damage were determined. Ostling and Johanson (1984), have increased the sensitivity in the detection of DNA damage in cells by improving the Microgel Electrophoresis techniques. The technique proposed by Ostling and Johanson is based on spiral breaks in DNA and the mechanism by which this structure can be pulled out and attenuated by electrophoresis. Singh et al. (1988) have developed a method that allows detection of DNA damage under alkaline conditions by modifying the microgel electrophoresis technique. In this technique, electrophoresis is carried out at high pH.

Table 1. Types of DNA damage that can be detected by Comet assay at different pH

N/N (pH: 7-8)	A/N pH: 10-12	A/A pH > 13
Double strand breaks	Double strand breaks	Double strand breaks
Cross links	Single strand breaks	Single strand breaks
	Excision repair sites	Excision repair sites
	Cross links	Cross links
		Alkali injuries

Versions such as N/N (Neutral Dissolution/Neutral Electrophoresis), A/N (Alkaline Dissolution/Neutral Electrophoresis) and A/A (Alkaline Dissolution/Alkaline Electrophoresis, pH 13) developed for different pHs (Table 1) are also used in the Comet assays (Gichner and Plewa, 1998; Lin et al., 2007).

After the neutralization step, the slides are stained and analyzed within a maximum of 6 hours. If analysis can not be done immediately, the preparations are kept in cold water for 10-15 minutes, then dehydrated in 100% ethanol and left to dry. The slides can be stored for a few months under appropriate conditions (Kocyigit et al., 2005, Gichner et al., 2008 a, b).

The comet formation of nuclei followed by electrophoresis enables quantification of DNA in comet tails after staining with an appropriate fluorochrome such as propidium iodide or ethidium bromide (Olive and Banath, 2006). Despite the use of sensitive but quite expensive dyes such as YOYO-1 (benzoxazolium-4-quinolinone oxazole yellow homodimer), DAPI (4,6-diamidino-2-phenylindole) in the detection of lower levels of DNA damage, it is reported that the efficiency of the cheaper etidium bromide was not different from other dyes mentioned (Gichner et al. 2006).

Following staining, migration of DNA from the nucleus is observed in damaged cells. Each damaged cell; a brilliant fluorescence head and length and fluorescence intensity has a comet image consisting of a tail connected to the number of DNA strand breaks in the cell. Undamaged cells appear as untailed, undisturbed nuclei. DNA images emerging at varying degrees of damage from circular to comet-like form are examined under a fluorescent microscope.

Utilizing the feature that the amount of DNA in a particular region is directly proportional to the fluorescence intensity in that region, and in parallel with the progress in digital imaging systems and analysis software programs, more accurate quantification methods have been developed. Image analysis systems can measure various comet parameters such as head length of damaged cells, percentage of DNA in head and tail, tail length and tail moment (Tice ve ark., 2000; Hartmann ve ark., 2003).

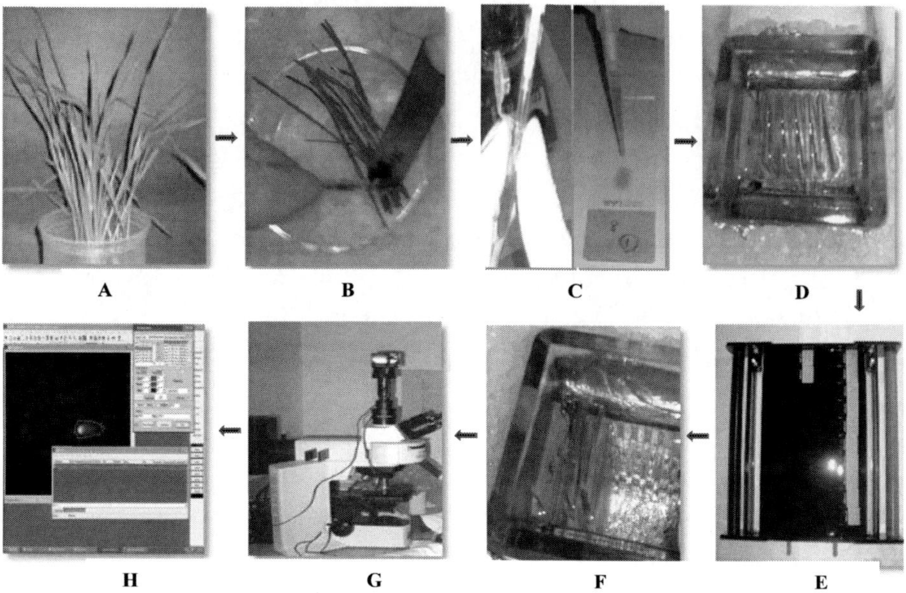

Figure 1. General illustration of the Comet assay. A, B: Isolation of nuclei from plant leaf tissue, C: The mix of isolated nuclei and LMP agarose transfer onto the slides, D: Unwinding of the slides in fresh alkaline or neutral electrophoresis buffer, E: Electrophoresis of the slides in the alkaline or neutral buffer, F: Neutralize the slides, G: Stain the slides with a DNA binding dye and view under fluroscence microscope, H: Analyse the results (Akan, 2012).

According to its head and tail shape of comet, the migration of DNA could be classified from 0 (no tail) to 4 (almost all DNA in tail) by visually (Kobayashi et al., 1995; Gichner et al., 2003; Kocyigit et al., 2005).

The use of Comet test with different methods affects the results. For instance, electrophoresis conditions (duration, applied voltage), lysis solution content (salt concentration and pH) and duration of application affect the sensitivity of the method. For this reason, optimizing the technique in comet labs is an important step (Hartmann et al., 2003).

COMET ASSAY IN PLANTS

Comet is an effective tool that can be applied easily to fungi, algae, plants, sea creatures, insects, vertebrate animals and humans, and to give information on environmental health (Gichner and Plewa 1998, Angelis et al., 2000; Menke et al., 2001; Dhawan et al., 2009). Although almost all plants are available for the Comet test, they are potatoes (Solanum tuberosum), tobacco (*Nicotiana tabacum*), soya bean (*Vicia faba*), onion (*Allium cepa*) and *Arabidopsis thaliana* in terms of ease of cultivation and cell isolation (Gichner et al., 2006; Mancini et al., 2006; Lin et al., 2008). Gichner and Plewa (1998) found that after exposure to *N. tabacum* plants with 1-8 mM Ethyl Methane Sulphonate (EMS), the highest amount of DNA damage was found in the plants exposed to the highest EMS concentration. Menke et al. (2000) investigated that the N-methyl-N-nitroso Urea (MNU) was genotoxic effective agent in *V. faba* plants in comparison with three different comet technique protocols (neutral/neutral, alkaline/neutral, alkali/alkaline). Gichner et al. (2008) conducted that comet analysis under alkaline conditions to determine the increase in induced DNA damage in *S. tuberosum* L. plant treated with EMS, Cd^{+2} and γ-rays, showed that the best dose-response curve was detected with the alkaline Comet technique.

SCGE, in combination with detection of Reactive Oxygen Species (ROS) and expression analysis of both DNA repair and antioxidant genes, can be used for the plant genotypes (Ventura et al., 2013). The accumulation of reactive oxygen species (ROS) are damage the cellular components such as DNA, proteins and lipids (Lopez et al., 2006).

The overproduction of ROS induced by some heavy metals can lead to the DNA damage in plants, animals and microorganisms living in aerobic conditions (Gichner, 2003; Mittler, 2002; Mancini et al., 2005), are the early responses of plant to heavy metals (Mittler, 2002; Hartley-Whitaker et al., 2001). Lin et al. (2007) reported that the enhancement in DNA damages was correlated to oxidative stress induced by Cd in leaves of *Vicia faba*, indicating that ROS accumulation in tissues caused subsequently DNA damage.

Table 2. Selected Comet assay studies in plant toxicology (Santos et al., 2015)

Stress	Species	Tissue	Comet type
Radiation			
Light	*A. thaliana*	Leaves	A/ApH > 13
UV	*A. thaliana*	Leaves	A/ApH > 13
X-ray	*A. thaliana*	Leaves	A/ApH > 13
Metals			
Monovalent	*V. faba*	Roots, leaves	A/ApH > 13
Divalent	*A. cepa*	Roots	A/ApH > 13
Trivalent	*A. thaliana*	Roots	N/N pH 8.4
Oxoanions	*A. cepa*	Roots	A/ApH > 13
Nanocompounds			
MWCNT	*A. cepa*	Roots	A/ApH > 13
Metal NPs	*A. cepa*	Roots, leaves	A/ApH > 13
Metal oxideNPs	*A. cepa*	Roots	A/A pH > 12.3
Quantum dots	*M. sativa*	Cell suspension	A/ApH > 13
Organic pollutants			
Dyes	*A. cepa*	Roots	A/ApH > 13
Pesticides	*A. cepa*	Roots	A/ApH > 13
Polyhalogenated	*A. cepa*	Roots	A/ApH > 13
Phytocompounds			
	A. cepa	Roots	A/ApH > 13
Others			
Osmostressors	*A. thaliana*	Seedlings	N/N pH 8
Control mutagens			
	A. cepa	Roots, leaves	A/ApH > 13

Comet assay has been used in different studies by use of plants particularly in environmental risk assessment (Table 2). Despite the fact that almost all plants are available for comet analysis, studies on *S. tuberosum, N. tabacum, V. faba, A.cepa* and *A. thaliana* have been found more widely in the literature (Gichner et al. 2006; Mancini et al., 2006; Lin et al., 2008).

Programmed Cell Death (PCD) is genetically controlled death and mediated by specific proteases and nucleases. During PCD cells are executed in a biological process and characterized by cell shrinkage, vacuolization, nucleus deformation and DNA fragmentation. In detailed studies of PCD DNA fragmentation is accepted as the most apparent indicator of PCD. DNA fragmentation in dying cells is visualized/measured by several analyses such as DAPI staining, TUNEL reaction, flow cytometry, gel electrophoresis and Comet assay (Vardar et al., 2011; 2012; 2015; 2016). Although the DNA damage can be visualized and measured by Comet assay, the whole of comets may not be considered as PCD. It is the fact that the damaged cells will subsequently be undergone irreversible PCD. Considering the fact that DNA damage might be repairable, the Comet tail can be no longer seen under proper conditions. Besides damaged cells die and disappear, but in comet assay the damaged cells are still present and detectable. Moreover, DNA fragmentation during PCD is characterized by DNA cleavage into 180-200 bp length oligonucleosomal fragments. During electrophoresis in Comet assay such small parts of DNA will probably disappear (Collins et al., 1995; Collins, 2004). To overcome these difficulties Singh (2000) described a visualization method of dying cells. In this technique cells are embedded in the agarose and lysed. But the researcher precipitated the DNA, instead of electrophoresis. The dying cells distinguished by halo of granular DNA and a hazy outer boundary (Singh, 2000). The inverse relationship between the application time and the DNA damage can be explained by the partition of the enzymatic processes in the formation of DNA damage of pollutants causing toxicity (Ali and Kumar, 2008). Akan (2012) reported that CPF application in barley plant demostrated that there was an inverse relationship between DNA damage and application time. Long-term studies are required to examine the repair of DNA damage, protective mechanisms and enzymatic processes is suggested in this study.

Agronomic crops as well as other plants are deliberately exposed to pesticides and other chemicals applied in modern agriculture. These pollutants are hazardous for the plants. Basudine and antracol pesticides, destroys crop and livestock. In our study, genotoxic potential of Basudine and Antracol were determined using Single Cell Gel Electrophoresis (SCGE) in tobacco (*Nicotiana tabacum* L.) which is model plant and it is suitable for genetic and physiological studies. The Comet assay has been widely used in the fields of genetic toxicology and environmental biomonitoring by evaluating DNA strand breaks.

Pesticide applications can cause damage due to many factors. These factors include the form of dose (how much is applied), duration (how long it is applied), and how contact occurs. In addition, the presence of any other chemicals exposed, the age and other characteristics of the organism must be taken into consideration.

In the study conducted by Aydın et al. (unpublish data), possible genotoxic effects of Basudin (fungicide - active ingredient: Diazinon) and Antrakol (insecticide - active ingredient: Probinep) pesticides on model organism tobacco plant (*Nicotiana tabacum* L.) were evaluated by single cell gel electrophoresis. Diazinon can reduce its fixation ability by nitrogenous soil organisms, an important plant nutrient and nitrogen fixation is a process that transforms atmospheric nitrogen into a form that can be used by plants. Diazinon applied to the soil is absorbed by plant roots and can be translocated in plants (Kamrin, M. A., 1997). Their half-life ranges from 2 to 14 days. Low temperature and high fat content increase the durability of diazinon in plants (Kamrin, M. A., 1997). Diazinon also causes genetic damage in plants. Diazinon application with onion (Grover et al., 1987) and barley (Kour and Grover, 1985a, Grover et al., 1987) resulted in abnormal chromosomal occurrences. In addition, seeds obtained from diazinon-treated plants contain mutations that reduce the proportion of chlorophyll in cells following germination (Grover et al., 1987). Diazinon applications to barley seeds or seedlings cause abnormal chromosomes to form in pollen cells (Kour and Grover, 1985b). Antrocol is used to destroy fungi causing damage to various vegetables and fruits (tomatoes, potatoes, onions, grapes, melons, etc.).

In this study, Basudin and Antrocol pesticides are used extensively controlling of pests on park and garden especially in İstanbul, Turkey. The genotoxic effects of these pesticides are not completely understood and the Comet assay was used for detecting DNA damage by these pesticides exposed to the model organisms, tobacco (*N. tabacum*).

Thus, the aim of this study was to evaluate the genotoxic effects of acute exposures (24, 48, 72 and 96h) to 0.925, 1.85, 2.775 mg/ml of Basudin and 0.5, 1, 1.5 mg/ml of Antracol on two-week-old tobacco seedling. Tobacco plants germinated in MS (Murashige and Skoog, 1962) medium and nucleus isolated from their leaves. Leaf samples from non pestiside treated tobacco seedlings grown on MS medium were used as negative controls, and leaf samples from seedlings treated with distilled water containing 0.1% H_2O_2 (during 24 hours) were used as positive controls. Leaf tissues were disrupted with sharp razors adding cold 0.4 M Tris- HCl buffer (pH7.5). The tissues were gently sliced and isolated for collection of nuclei in the buffer. The mixture was centrifuged to remove the debris. All steps were performed in dim light and ice to prevent induction of DNA damage. Nuclei suspension was mixed with the liquid Low Melting Agarose (LMA) (1:1) by gently pipetting and as soon as agarose/nuclei mixture were spread on the slide that previously coated with High Melting Agarose (HMA). The slides were incubated in electrophoresis buffer to facilitate DNA unwinding. Thereafter, the slides were placed on horizontal electrophoresis tank filled with fresh, cold electrophoresis buffer kept cold using ice followed by performing electrophoresis. After neutralization each slide was stained with ethidium bromide and kept at 4°C until analysis. These slides observed under an epifluorescence microscope (BX51TF, Olympus) equipped with a CCD camera (Canon A640) and fifty cells per slide were analyzed with the Comet image analysis system (BAB BS200ProP). For each sample, 50 random cells were scored. The % tail DNA and olive tail momentum parameters were calculated as a measure of DNA damage. % Tail DNA values were shown as the percentage of DNA in comet tail. Olive tail moment values were shown as multiplication of % tail DNA and difference between center of gravity of DNA in the tail and center of gravity of the head DNA. The results showed that Basudin and Antrocol produce genotoxic damage in leaf nucleus of *N. tabacum*. The comet scores obtained for tobacco after all time points of exposure to Basudin and Antrocol were significantly higher than the respective negative control ($P < 0.001$). In conclusion the result of this work showed that Basudin and

Antrocol induced genotoxic effects on the plant species *N. tabacum.* The genotoxic effects were observed in all Basudine and Antracol concentrations and exposure times in tobacco (Table 3, Figure 2 and Figure 3).

It has been determined that diazinon inhibits the activity of an enzyme in soil microorganisms that "plays an important role" in the nitrogen cycle (Frankenberger and Tabatabai, 1991). Diazinon also reduces the amount of nitrogen compounds in the soil (Ingham and Coleman, 1984).

Comet analysis of the findings, as well as the determination of pesticide spraying and harvesting time of use to determine appropriate doses contributed to the work.

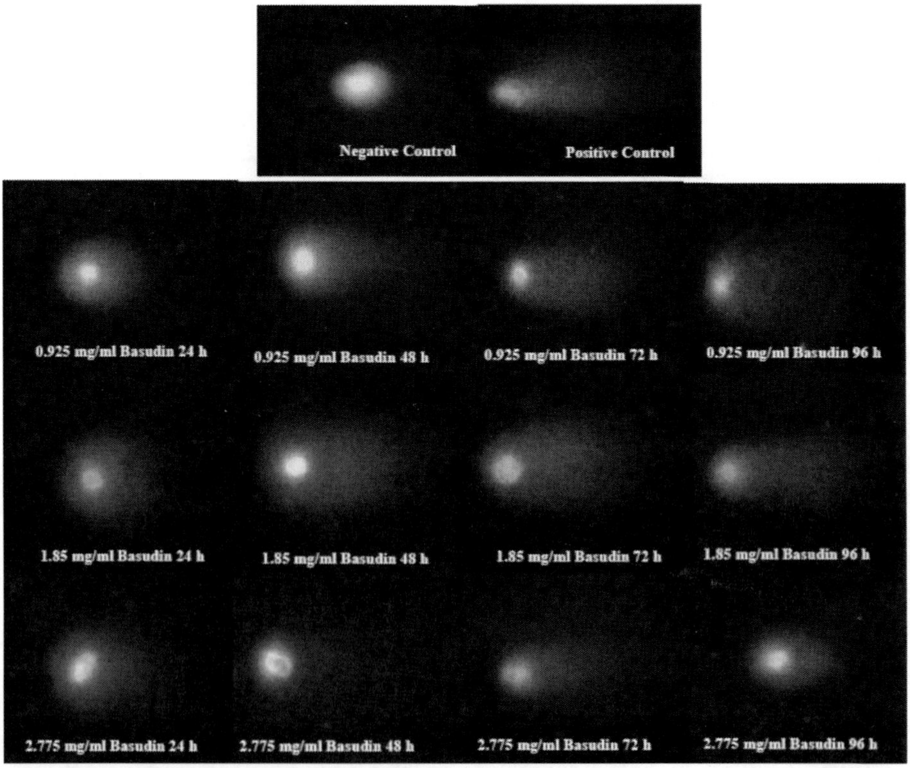

Figure 2. DNA migration pattern from *Nicotiana tabaccum* L., exposed to different concentrations of Basudin with different exposure times. Objective: 20X. Dye: EtBr (Unpublised data).

Table 3. DNA damage at different concentrations and exposure times of Basudin and Antracol (%DNA tail and Olive tail moment)

Pesticide	Dose	Time (hour)	%DNA tail	Olive tail moment
Basudin	(-) control		26	0.26
	(+) control	18	36	0.26
	0.925 mg/ml	24	29	0.20
		48	28	0.21
		72	38	0.29
		96	36	0.34
	1.85 mg/ml	24	33	0.20
		48	27	0.12
		72	29	0.16
		96	23	0.17
	2.775 mg/ml	24	18	0.17
		48	30	0.18
		72	23	0.21
		96	31	0.15
Antracol	(-) control		26	0.26
	(+) control	18	36	0.26
	0.5 mg/ml	24	26	0.15
		48	28	0.17
		72	27	0.15
		96	23	0.13
	1 mg/ml	24	20	0.12
		48	17	0.14
		72	32	0.16
		96	20	0.16
	1.5 mg/ml	24	26	0.15
		48	33	0.22
		72	28	0.12
		96	25	0.11

In vitro culture factors could apparently lead to oxidative stress and genotoxicity. The plants produce oxine and cytokine in their own tissues. These hormones can reach high doses in plants by supplementing these plant growth regulators from outside in plant tissue culture studies. The production of auxin and cytokine may be distinct in plants for different gene sources. In recent years, in vitro techniques have ben extensively used for abiotic stress (Şen, 2012) and genotoxic effects.

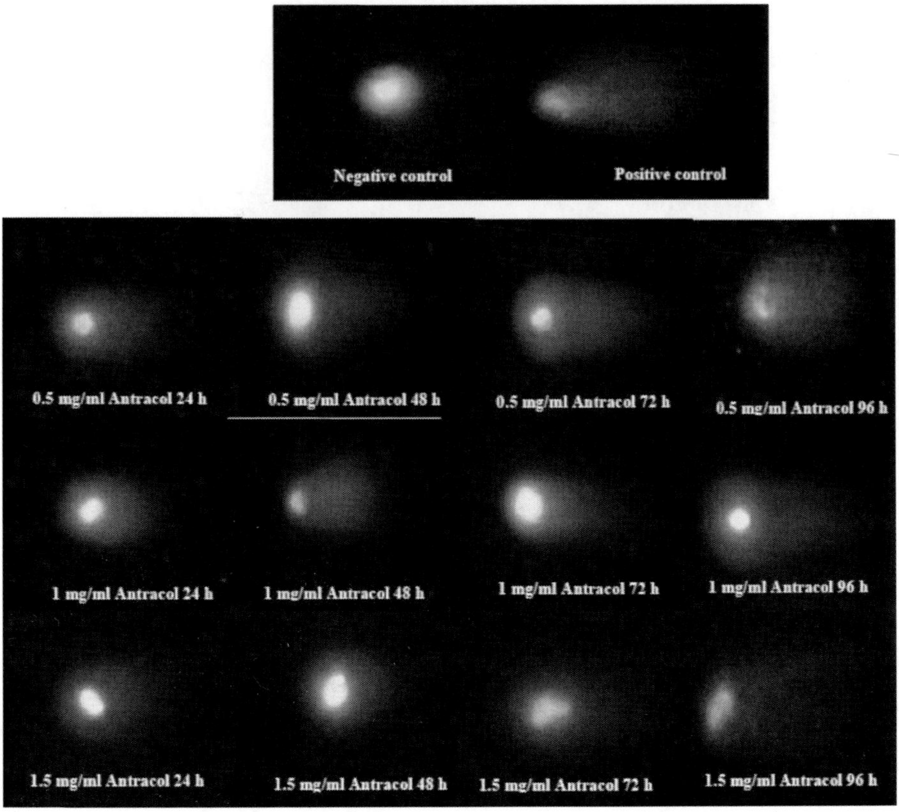

Figure 3. DNA migration pattern from *Nicotiana tabaccum* L., exposed to different concentrations of Antracol with different exposure times. Objective: 20X. Dye: EtBr (Unpublised data).

In another study conducted by our group, it was evaluated the changes on genotoxicity derived from androgenic callus tissues with the effect of plant growth regulators, genotypes and pretreatments in sunflower by Comet assay. The method of Çavaş and Könen (2007) was used by some modifications in this study. Obtaining callus tissues were used as material to detect the DNA damage levels by Comet assay. Leaf of sunflower grown in MS medium for two weeks was used as a negative control and leaf of sunflower that grown in MS medium supplemented with 150 μm $AlCl_3$ for two weeks was used as positive control. Callus and leaf tissues were disrupted with sharp razors adding cold 0.4 M Tris- HCl buffer (pH=7.5). The slides were analyzed under a fluorescent microscope at 20X magnification. For each sample, 50 random cells were scored. Comet parameters were measured and analyzed using KAMERAM software. The % tail DNA and olive tail momentum parameters were calculated as a measure of DNA damage. % Tail DNA values were shown as the percentage of DNA in comet tail. Olive tail moment values were shown as multiplication of % tail DNA and difference between center of gravity of DNA in the tail and center of gravity of the head DNA. Overall results we obtained indicated that genotype, medium type and pretreatment factors especially plant growth regulators could apparently lead to genotoxicity and medium type is the most effective parameter for genotoxic effect on sunflower (unpublished data).

In conclusion; Comet assay is a fast, easy and reliable method to evaluate DNA damage on living organisms and in vitro and in vivo studies with different concentration and exposure periods induced by various types of genotoxic agents. Comet assay also can be employed in specific tissues or cells such as roots and leaves. These characteristics of the assay make it to be used in evaluating various types of environmental toxic conditions. Although Comet asay is most used the evaluating the acute effects of genotoxic agents, it can be also used to reveal DNA repair in plant physiology, as this technique may help elucidating pathways involved in plant development, cell cycle/programmed cell death, or even plant disease resistance. Long-term genotoxicity studies can be an important approach to gaining insight into the ability of organisms to repair DNA and other protective mechanisms.

ACKNOWLEDGMENT

This work was supported by the Research Foundation of Marmara University (BAPKO no. FEN-B-090909-0293).

REFERENCES

Akan, T. (2012). *Investigation of in vitro genotoxic effects of organophosphorus pesticides in barley (Hordeum vulgare L.) and zebrafish (danio rerio).* Master thesis.

Ali, D. and Kumar, S. (2008). Long-term genotoxic effects of monocrotophos in diffrent tissues of freshwater fish *Channa punctatus* (Bloch) using alkaline single cell gel electrophoresis. *Science of Total Environment,* 405, 345-350.

Angelis, K.J., McGuffie, M., Menke, M., Schubert, I. (2000). Adaptation to alkylation damage in DNA measured by the Comet assay. *Environmental and Molecular Mutagenesis,* 36: 146–50.

Collins, A.R., Ma, A-G., and Duthie, S.J. (1995). The kinetics of repair of oxidative DNA damage (strand breaks and oxidised pyrimidines) in human cells. *Mutat. Res.* 336, 69–77.

Collins, A.R. (2004). The Comet Assay for DNA Damage and Repair. *Molecular Biotechnology.* 26, 249-261.

Çavas, T., Könen, S. (2007). Detection of cytogenetic and DNA damage in peripheral erythrocytes of goldfish (*Carassius auratus*) exposed to a glyphosate formulation using the micronucleus test and the comet assay. *Mutagenesis.* 22 (4), pp. 263–268.

Dhawan, A., Bajpayee, M., Parmar, D. (2009). Comet assay: a reliable tool for the assessment of DNA damage in different models. *Cell Biology Toxicology.* 25, 5-32.

Gichner, T., Plewa, M.J. (1998). Induction of somatic DNA damage as measured by single cell gel electrophoresis and point mutation in leaves of tobacco plants. *Mutation Research,* 401, 143-152.

Gichner, T. (2003). Differential genotoxicity of ethyl methanesulphonate, N-ethyl-N-nitrosourea and maleic hydrazide in tobacco seedlings based on data of the Comet assay and two recombination assays. *Mutat. Res.* 538, 171-179.

Gichner, T., Patkova, Z., Kim, J.K. (2003). DNA damage measured by the Comet assay in eight agronomic plants. *Biol. Plant.* 47, 185-188.

Gichner, T., Patkova, Z., Szakova, J., Demmerova, K. (2006). Toxicity and DNA damage in tobacco plants growing on soil polluted with heavy metals. *Ecotox. Environ. Safe.* 65, 420-426.

Gichner, T., Znidar, I., Szakova, J. (2008a). Evaluation of DNA damage and mutagenicity induced by lead in tobacco plants. *Mutat. Res.* 652, 186- 190.

Gichner, T., Patkova, Z., Szakova, J., Znidar, I., Mukherjee, A. (2008b). DNA damage in potato plants induced by cadmium, ethyl methanesulfonate and γ-rays. *Environmental and Experimental Botany,* 62,113-119.

Grover, I.S., Dhingra, A.K., Adhikari, N., Lidhar, S.S. (1988). Genotoxic effects of pesticides. *Cell Bio and Toxicol.* 4(4), 519.

Hartley-Whitaker, J., Ainsworth, G., Meharg, A.A. (2001). Copperand arsenate-induced oxidative stress in *Holcus lanatus* L. clones with differential sensitivity[J]. *Plant, Cell and Environment.* 24, 713–722.

Hartmann, A., Agurell, E., Beevers, C., Brendler-Schwaab, S., Burlinson, B., Clay, P., Collins, A., Smith, A., Speit, G., Thybaud, V., Tice, R.R. (2003). Recommendations for conducting the *in vivo* alkaline Comet assay. *Mutagenesis.* 18, 45-51.

Ingham, E.R., Coleman D.C. (1984). Effects of streptomycin, cycloheximide, Fungizone, captan, carbofuran, Cygon, and PCNB on soil microorganisms. *Microb. Ecol.* 10:345-358.

Kamrin, M.A. (1997). *Pesticide Profiles: Toxicity, Environmental Impact and Fate;* Lewis Publishers: Boca Raton. 157-159.

Kour, P. and Grover, I.S. (1985a). Cytological effects of some organophosphorus pesticides. *I. Mitotic effects. Cytol.* 50:187-197.

Kour, P. and Grover, I.S. (1985b). Cytological effects of some organophosphorus pesticides. *II. Meiotic effects. Cytol.* 50:199-211.

Kobayashi, H., Sugiyama, C., Morikawa, Y., Hayashi, M., Sofuni, T. (1995). A comparison between manual microscopic analysis and computerized image analysis in the single cell gel electrophoresis assay. *MMS Commun.* 3,103-115.

Kocyigit, A., Keles, H., Selek, S., Guzel, S., Celik, H., Erel, O. (2005). Increased DNA damage and oxidative stress in patients with cutaneous leishmaniasis. *Mutat. Res.* 585, 71-78.

Lin, A., Zhang, X., Chen, M., Cao, Q. (2007). Oxidative stress and DNA damages induced by cadmium accumulation. *J. Environ. Sci.* 19, 596-602.

Lin, A., Zhang, X., Zhu, Y.G., Zhao, F.J. (2008). Arsenate-induced toxicity: effects on antioxidative enzymes and DNA damage in *Vicia faba. Environmental Toxicology and Chemistry.* 27 (2), 413-419.

Lopez E., Arce C., Oset-Gasque M.J. et al. (2006). Cadmium induces reactive oxygen species generation and lipid peroxidation in cortical neurons in culture [J]. *Free Radical Biology & Medicine.* 40, 940–951.

Mancini, A., Buschini, A., Restivo, F.M., Rossi, C., Poli, P. (2006). Oxidative stress as DNA damage in different transgenic tobacco plants. *Plant Sci.* 170, 845-852.

Menke, M., Chen Peng, I., Angelis, K.J., Schubert, I. (2001). DNA damage and repair in *Arabidopsis thaliana* as measured by the comet assay after treatment with different classes of genotoxins. *Mutat. Res.* 493, 87–93.

Mittler R. (2002). Oxidative stress, antioxidants and stress tolerance [J] Trends in Plant Science, 7: 405–410. Olive, P.L., Banath, J.P. (2006). The Comet assay: a method to measure DNA damage in individual cells. *Nat. Prot.* 1, 23-29.

Olive, P.L., Banath, J.P. (2006). The Comet assay: a method to measure DNA damage in individual cells. *Nat. Prot.* 1, 23-29.

Östling, O., Johanson, K.J. (1984). Microelectrophoretic study of radiationinduced DNA damages in individual mammalian cells. *Biochem. Biophys. Res. Commun.* 123, 291-298.

Rank, J., Syberg, K., Jensen, K. (2009). Comet assay on tetraploid yeast cells. *Mutation Research.* 673, 53-58.

Rydberg, B., Johanson, K.J. (1978). Estimation of single strand breaks in single mammalian cells. In: Hanawalt PC, Friedberg EC, Fox CF, editors. *DNA Repair mechanisms.* New York, NY: Academic. P. 465–8.

Santos, C.L.V., Pourrut, B. and Ferreirade Oliveira, J.M.P. (2015). The use of comet assay in plant toxicology: recent advances. *Front.Genet.* 6: 216.

Singh, N.P. (2000) A simple method for accurate estimation of apoptotic cells. *Exp. Cell Res.* 256, 328–337.

Şen, A. (2012). Oxidative stress studies in plant tissue culture. *Antioxidant Enzyme. 3*, 59-88.

Tatli, M.M., Minnet, C., Kocyigit, A., Karadag, A. (2008). Phototherapy increases DNA damage in lymphocytes of hyperbbilirubinemic nonates. *Mutat. Res.* 654 (1), 93-95.

Tice, R.R., Agurell, E., Anderson, D., Burlinson, B., Hartmann, A., Kobayashi, H. (2000). The single cell gel/comet assay: guidelines for *in vitro* and *in vivo* genetic toxicology testing. *Environ. Mol. Mutagen.* 35, 206-21.

Vardar, F., Akgül, N., Aytürk, Ö., Aydın, Y. (2015). Assessment of aluminum induced genotoxicity with comet assay in wheat, rye and triticale roots. *Fresenius Environmental Bulletin.* 37, 3352-3358.

Vardar, F., Çabuk, E., Aytürk, Ö., Aydın, Y. (2016). Determination of aluminum induced programmed cell death characterized by DNA fragmentation in Gramineae species. *Caryologia.* 69, 111-115.

Vardar, F., İsmailoğlu, I., İnan, D., Ünal, M. (2011). Determination of stress responses induced by aluminum in maize (*Zea mays* L. Karadeniz Yıldızı). *Acta Biologica Hungarica.* 62, 156-170.

Vardar, F., Ünal, M. (2012). Ultrastructural aspects and programmed cell death in the tapetal cells of *Lathyrus undulatus* Boiss. *Acta Biologica Hungarica.* 63, 56-70.

Ventura, L., Giovannini, A., Savio, M., Dona, M., Macovei, A., Buttafava, A., Carbonera, D., Balestrazzi, A. (2013). Single cell gel electrophoresis (Comet) assay with plants: research on DNA repair and ecogenotoxicity testing. *Chemosphere.* 92, 1-9.

In: Advances in Health and Natural Sciences ISBN: 978-1-53614-639-4
Editors: Burcu Yuksel et al.

Chapter 13

Evaluation of the Sensitivity of the Nurses Working in Surgical Services of a Public Hospital to Patient Rights

Selda Mert Boga[1,*]
and Aylin Aydin Sayilan[2]
[1]Kocaeli Vocational School of Health Services,
Kocaeli University, Kocaeli, Turkey
[2]Department of Nursing, Kırklareli University, Kırklareli, Turkey

Abstract

This study was designed as a descriptive study, to analyze the sensitivities of the nurses working in the surgical services of a state hospital on patient rights. The universe of the study consisted of 61 nurses working in the surgical services of a state hospital while the sample of the study consisted of 54 nurses volunteered to participate in the study between July 26 and October 1, 2010. The data of the study

[*] Corresponding Author E-mail: seldamertboga@gmail.com.

were collected with the personal information form and the Belief of Patient Rights Form (BPRF). In the statistical analysis of the results, frequency, percentage, mean, standard deviation, Mann Whitney U and Kruskal Wallis-H tests were used. The mean total score of BFRF of the nurses was 32.85 ± 6.66; the differences between the nurses' scores on the BFRF and the subscales of privacy and private life and application were significant according to their ages ($p < 0.05$); the difference between the nurses' scores on the consent subscale was significant according to their education levels ($p < 0.05$). It was determined that the surgical nurses who were 29 years old or younger, university graduates, had less than 5 years' study period and took courses on patient rights, had high sensitivity to patients' rights. It can be recommended to add related courses in the curriculum during the education and to plan training programs by considering age group, the level of education, working time and the status of taking related courses, for improving nurses' sensitivity to patient rights.

Keywords: surgical nursing, surgical service, hospital, awareness, patients' rights

INTRODUCTION

Patient rights are defined as the rules of conduct between people who benefit from health services and the health institutions and personnel who provide them (Teke et al., 2007; Ozakar Akca et al., 2015). As these rights relate to particular circumstances in human life, it is also defined as third generation human rights and covers the application of human rights and values in health services. Top priorities of patient rights include: to be respected as a human being, receive health service in a high level, get information, get confirmation for medical processes, privacy, and respect for private life, while receiving care and treatment continuity (Erbil, 2009; Ozakar Akca et al., 2015).

In recent years, advances in science and technology have made being informed about treatment and care initiatives necessary for patients (Savaskan, 2006). This change and development has led the nurse's care function to mainly focus on new roles. One of the contemporary roles of the nurse is to stand up for patient rights (Ardahan, 2003). In this context, nursing is aimed to be a discipline protecting patient rights, implementing care and treatment, ensuring the safety of the patient, advocating and implementing the right thing for the patient (Ardahan, 2003; Ozer et al., 2009; Erzincanli & Zaybak, 2015). Therefore, nurses have obtained the identity of being a health professional being a guide of the patients to know and acquire their rights. Patients who cannot choose the best for themselves, often do not have knowledge of their rights therefore they need the help of a nurse who can guide them on how to obtain these rights. Nurses have more interactions with patients than other health care team members. For this reason, it is very important that nurses have enough information about the subject and can apply this knowledge in practice (Teke et al., 2007, Zaybak et al., 2012, Ozakar-Akca et al., 2015, Oztas & İyigun, 2016). The sensitivity of nurses to patient rights also positively affects the satisfaction levels of patients in the institution (Tanriverdi & Ozmen 2011; Erzincanli & Zaybak, 2015).

It is the ethical responsibility of nurses to undertake the role of patient advocacy in this period because the individuals/patients undergoing surgical intervention experience anesthesia and sedation. In surgical departments in which invasive procedures are mostly performed, it is also important to obtain the patient's informed consent, the nurse has the ethical responsibility to obtain the informed consent after understanding of the operation by the patient (Karaoz, 2000; Temiz et al., 2017). In the study of Ozer et al., (2009) which investigated the knowledge levels of patients treated in surgical clinics, it was reported that the mean knowledge scores of the patients was 11.67 ± 3.32 on a 17 point scale, 86.1% of the patients did not receive information about patient rights during hospitalization, the patients stated that the highest importance was given to the right of having a hospital attendant and while the lowest importance was given to the right of requesting information about their health status verbally and by writing in the services in which they stayed.

Within the scope of ethical principles, securing confidentiality of patients is also among the responsibilities of nurses when patients are not able to protect their own privacy due to their state of consciousness during the surgical preparation period, during the procedure, and in the postanesthesia care unit (Karaoz, 2000; Tumer et al., 2011; 2017). It was determined that most of the studies related to the subject analyzed the level of knowledge of the medical staff (Teke et al., 2007; Zincir & Kaya 2009; Kavak et al., 2014; Khalaf et al., 2014; Ozakar-Akca et al., 2015) and the patients (Ozer et al., 2012; Erzincanli & Zaybak 2015; Oztas & İyigun 2016) about patient rights. Therefore, this study was designed and carried out in order to determine the sensitivity of the nurses working in surgical units to patient rights because of lack of the studies on the sensitivity of nurses working in surgical units about patient rights. For this purpose, three questions below were asked

- What are the knowledge levels of the surgical nurses about patient rights?
- How is the surgical nurses' sensitivity to patient rights?
- Is there any difference between personal and occupational characteristics of surgical nurses and sensitivity to patient rights?

MATERIALS AND METHODS

Universe and Sample of the Study

The study was planned as a descriptive and cross-sectional study and was carried out between July-October 2010. The universe of the study consisted of 61 nurses working in the surgical clinics of a state university. No sample selection criteria were used, 54 nurses who agreed to participate in the study were included in the study.

Data Collection

The research data were collected with the personal information form and the Belief of Patient Rights Form (BPRF). In the personal information form, the information such as age, gender, marital status, education status, duration of study, education about patient rights, meeting with the patient rights concept, having a problem related with patient rights and the presence of a unit related to patient rights in the institution where the nurse works are included.

The BRPF is a 5-point liker-type scale that consists of 6 sub-dimensions (human rights and values in health care services, informing, approval, privacy and private life and application) and 18 questions determining the views of nurses on their beliefs about patient rights. The participants in the study were asked to mark one of the following expressions: "absolutely agree", "I agree", "I slightly agree", "I disagree", "I absolutely disagree". In the evaluation of the form, 1 point was given for the answer "Absolutely Agree", 2 points was given for the answer "Agree", 3 points was given for the answer "Slightly Agree", 4 points was given for the answer "Disagree" and 5 points was given for the answer "Absolutely disagree". The lowest possible score is 18 while the highest possible score is 90. Low scores indicate high agreement on patient rights (high sensitivity), and high scores indicate low agreement on patient rights (low sensitivity).

Statistical Analysis

The data were analyzed by SPSS 20 package program. Mann Whitney U and Kruskal Wallis-H tests were used because the data did not fit normal distribution. $p < 0.05$ was considered as statistically significant.

Ethical Statement

The written permission was obtained from the Hospital Directorate. Participants were informed about the aim of the study; A verbal permission was obtained from each participant who voluntarily participated in the study. Participants were reassured about the confidentiality of their information; they were informed about the aim and scope of the study.

Limits of the Study

The study was conducted with the surgical nurses in a state hospital, this may be accepted as a limitedness of the study in terms of generalizability.

RESULTS AND DISCUSSION

The study was conducted to determine the sensitivity of surgical nurses to patient rights; the results were interpreted and discussed below. No study was found to analyze the sensitivity of surgical nurses to patient rights as a result of that some difficulties were experienced in the discussion of our findings.

It was found that 40.7% of the nurses were between the ages of 30-34, 94.4% of them were female; 72.2% of them were married; 51.9% of them had an associate's degree; 31.5% of them started tow work 11-15 years before. The findings on the personal and professional characteristics of the nurses were similar to the findings of Zincir & Kaya (2009), Kavak et al., (2014).

Table 2 shows where the nurses first met with the concept of patient rights, the status for taking a training about patient rights and having a problem about patient rights and the presence of the patient rights unit at the hospital. Accordingly, it was determined that 55.6% of the participant nurses met the concept of patient rights for the first time in the hospital. The patient rights need to be known and implemented by every health care staff in order to increase the quality of the service provided in the field of health, to prevent patient rights violations both morally and legally and to increase the satisfaction level of the patients by meeting their expectations

(Tanrıverdi &, Ozmen 2011; Oztas &, İyigun 2016). Similarly, Ozakar-Akca et al., (2015) also found that 75.9% of nurses heard the concept of patient rights in the hospital for the first time; Teke et al., (2007) found that 64.2% of nurses met patient rights by using out of school sources. This condition suggests that nurses who are very close to the patients and who spend most of their daily working time with the patients do not learn enough information about patient rights in school education and practices.

It was determined that 77.8% of the participant nurses received education on patient rights. There is a close relationship between the essential values that constitute the nursing profession (Karaoz, 2000; Acaroglu, 2014) and the values that constitute patient rights (Patient Rights Regulation 1998). For providing ethical based care, the required importance should be given to teach care philosophy and care behaviors to them during their education period.

According to Karaoz (2000), Kurtz & Wang (1991), educators, researchers and managers should collaborate in order to achieve this goal. 48.7% of the nurses got education on patient rights in the study of Savaskan (2006); 55% of the nurses got education on patient rights in the study of Teke et al., (2007); 47.2% of the nurses got education on patient rights in the study of Kavak et al., (2014); 92.3% of the nurses got education on patient rights in the study of Ozakar Akca et al., (2015). The studies have shown that health staff generally have information about the application of patient rights. According to the findings of our study and other studies (Teke et al., 2007; Kavak et al., 2014), it can be stated that the majority of nurses who got education on patient rights did not constitute the majority of educated nurses; more importance should be given to the education about patient rights before and after the education. In the study, it was determined that 68.5% of the nurses did not encounter a problem about patient rights while 31.5% of them encountered a problem.

Table 1. The frequency distribution related with the personal and Professional characteristics of the nurses (n = 54)

Characteristics		n	%
Age	≤29	17	31,5
	30-34	22	40,7
	≥35	15	27,8
Gender	Female	51	94,4
	Male	3	5,6
Marital status	Married	39	72,2
	Single	15	27,8
Educational status	HS/VHS	16	29,6
	Associate Degree	28	51,9
	Bachelor	10	18,5
Working time	0-5 years	15	27,8
	6-10 years	13	24,1
	11-15 years	17	31,5
	≥16 years	9	16,6
Total		54	100,0

HS: High School; VHS: Vocational High School.

Table 2. Distribution of the nurses' knowledge on patient rights (n = 54)

Knowledge about patient rights		n	%
The place for meeting the concept of patient rights for the first time	Not meet	1	1,9
	Book, brochure	2	3,6
	TV	3	5,6
	Symposium, conference	6	11,1
	School	12	22,2
	Hospital	30	55,6
The status for getting training on patient rights	Yes	42	77,8
	No	12	22,2
The status for experiencing a problem related with patient rights	Yes	17	31,5
	No	37	68,5
The presence of the patient rights unit in the hospital	Yes	50	92,6
	No	4	7,4
Total		54	100,0

It is known that patients undergo extreme stress and need help from someone else to cope with this stress, from hospital application to discharge. Nurses are the most important health staff that can help patients by giving explanatory information about what to do in the treatment process. In this context, the nurses fulfill their own responsibilities by protecting the rights of the patient or by helping the protection of patient rights (Teke et al., 2007, Ozakar-Akca et al., Similarly, Ozakar-Akca et al., (2015) found that 43.8% of the nurses encountered a problem about patient rights while Teke et al., (2007) found that 18.3% of them encountered a problem about patient rights. It is remarkable that nurses who have the most interaction with patients in healthcare and who have an important role in advocating patient rights, encounter problems with patient rights at the rate of 18% to 44%. Patient rights are the rights that hospital management, patients and health professionals should protect together for a healthier care environment (Karaoz, 2000; Ozer et al., 2009; Erzincanlı &Zaybak 2015). We suggest that the intensification of educational programs, including patients-patient relatives and healthcare staff, will result in a positive change in behavior for resolving the problems related to patient rights and ensuring a healthier, more satisfying care environment. The patient rights related problems being encountered by nurses in practice is a different research subject.

In this study, it was determined that 92.6% of the nurses stated that there was a patient rights unit in the hospital. The Patient Rights Unit is a unit that is responsible for the training of health staff and patients about patients 'rights (Patients' Rights Regulation 1998), which accepts applications to prevent violations related with patient rights, provides solutions to urgent problems and informs a higher unit. Contrarily, 65.2% of the nurses stated that they did not have a board protecting the patient rights in the institution while 26.6% of the nurses stated that there was no information about the presence of the institution according to the findings of Savaşkan (2007). High awareness of the nurses about the presence of the patient rights unit in the hospital is a positive result for us.

Table 3. Distribution of the scores of the nurses on the overall BRPF scale and its subscales

Subscales	n	Mean	SD	Med	Min - Max
Human rights and values in health services	54	4,33	1,39	4	3 - 7
Informing	54	4,94	1,47	5	3 - 8
Consent	54	5,89	1,7	6	3 - 9
Privacy and private life	54	4,41	1,22	4	3 - 6
Care and treatment	54	6,13	1,52	6	3 - 11
Application	54	7,15	1,95	7	3 - 12
Total	54	32,85	6,66	34	19 - 47

Med.: Median; Min.: Minimum; Max.: Maximum; SD: Standard Deviation.

Table 4. Distribution of the scores of the nurses on the overall BRPF scale and its subscales according to their ages

Subscales			**Age**				**KW H Test**	
		n	Mean	SD	Med	Min - Max	H	p
Human rights and values in health services	≤29 years	17	4,06	1,39	3	3 - 6	1,22	0,542
	30-34 years	22	4,36	1,33	4	3 - 6		
	≥35 years	15	4,6	1,5	5	3 - 7		
Informing	≤29 years	17	4,47	1,46	5	3 - 8	4,28	0,117
	30-34 years	22	5,0	1,6	6	3 - 8		
	≥35 years	15	5,4	1,18	6	3 - 8		
Consent	≤29 years	17	5,29	1,72	5	3 - 9	3,52	0,172
	30-34 years	22	6,14	1,78	6	3 - 9		
	≥35 years	15	6,2	1,47	6	3 - 8		
Privacy and private life	≤29 years	17	3,76	1,15	3	3 - 6	7,89	0,019*
	30-34 years	22	4,59	1,22	4	3 - 6		
	≥35 years	15	4,87	1,06	5	3 - 6		
Care and treatment	≤29 years	17	5,71	2,11	6	3 - 11	2,33	0,312
	30-34 years	22	6,27	1,28	6	4 - 9		
	≥35 years	15	6,4	0,91	6	5 - 8		
Application	≤29 years	17	6,12	2,18	5	3 - 10	6,84	0,033*
	30-34 years	22	7,55	1,41	8	5 - 10		
	≥35 years	15	7,73	2,02	8	5 - 12		
Total	≤29 years	17	29,41	7,44	26	19 - 47	7,18	0,028*
	30-34 years	22	33,91	6,18	34	24 - 44		
	≥35 years	15	35,2	5,05	36	22 - 42		

*p < 0,05; Med.: Median; Min.: Minimum; Max.: Maximum; SD: Standard Deviation; KW: Kruskal Wallis.

When the mean scores of the nurses on the overall BRPF and all subscales of it were analyzed (Table 3), the mean score of the nurses on the subscale of human rights and values in health services was 4.33 ± 1.39; the mean score of the nurses on the subscale of informing was 4.94 ± 1.47; the mean score of the nurses on the subscale of consent was 5.89 ± 1.7; the mean score of the nurses on the subscale of care and treatment was 6.13 ± 1.52; the mean score of the nurses on the subscale of application was 15 ± 1.95; the mean total score of the nurses on the BRPF scale was 32.85 ± 6.66. The scores can be obtained from the BRPF Scale varies between 18 and 90, therefore it can be said that the mean score of the nurses on the BRPF Scale was low. This result indicates hat the sensitivity of the nurses to patient rights was high and the result was at the desired level.

In Turkey, the most important regulation on patient rights is the Patient Right Regulation, it was published in the Turkish Official Gazette (No. 23420) at 1 August 1998. This ordinance is an important document that directly and perceptibly regulates patient rights in Turkey. In the Patient Rights Regulation, the rights such as "benefiting from health services, asking for information, choosing and changing the health institution, appropriate diagnosis in compliance with medical requirements, treatment, requesting respect for privacy, not being subjected to medical applications without consent, requesting confidentiality of information, refusing the treatment" (Patient Rights Regulation 1998: 23420). The result of this study related with the mean total score of the nurses on the BRPF Scale was similar to the results of Savaskan (2006); this result suggested that the nurses provided care for the essential principles that constitute patient rights.

In Table 4, there was not a statistically significant difference between the scores of the nurses on the 'human rights and values in health services', 'informing', 'consent', 'care and treatment' subscales according age groups ($p > 0.05$). There were statistically significant differences between the scores of the nurses on the 'privacy' and 'private life', application subscales and overall scale ($p < 0.05$). In general, the nurses with the ages of 29 years or younger had lower scores on the overall BRPF and all subscales; they had higher sensitivity to patient rights. The studies comparing the sensitivity of the nurses to patient rights according to age groups, have found different results. There was no statistically significant difference between the nurses' sensitivity to patient rights according to age groups ($p > 0.05$) in the study of Savaskan (2006) and Kavak et al., (2014). In the study of Zincir & Kaya (2009), it was reported some similar which are similar to our results, the mean knowledge level of the health workers with the ages of 29 or younger was significantly higher than the others in the older age group ($p < 0.05$). In addition, Khalaf et al., (2014) analyzed the knowledge and attitudes of health care staff about patient rights. It was found that slightly more than one third of the participants (36%) knew patient rights (40.4% of the doctors, 34.2% of the nurses); it was reported that the rights of having 'a good care and management' and 'privacy and confidentiality' were the best known rights. In this study, the nurses aged 29 years or younger had higher mean scores compared to older age group. The reason is thought to be the fact that the patient rights regulation that went in effect in 1998 and after that year this regulation started to be taught as a course in pre-graduation education. Moreover, it can be suggested that the nurses in the young age group are more likely to adopt these rights by considering the fact that the significant differences were found for the subscales of 'privacy and private life' and application'.

There was no statistically significantly difference but the scores of the nurses that with a bachelor's degree were lower on the overall BRPF and it's subscales as 'human rights and values in health services', 'informing', 'privacy and private life', 'care and treatment', 'application' subscales compared to the nurses with an associate's degree ($p > 0.05$). A statistically significant difference was found between the scores of the nurses on the consent subscale according to their educational status ($p < 0.05$, Table 5). It was observed that the mean score of the nurses having a bachelor degree on the 'consent' subscale was significantly lower compared to the nurses having an associate's degree. The nurses with a bachelor degree were generally more sensitive to patient rights. It is a legal and ethical obligation to obtain verbal and written consent from the patient or his/her relatives prior to the initiation of any surgical procedure to be applied to the patient (the Patient Rights Regulation 1998). In the study of Kavak et al., (2014), there was a statistically significant difference between the scores of the nurses on the consent subscale according to their education status ($p < 0.05$); it was found that the nurses who graduated from a health vocational high school had significantly higher knowledge level on patient consent. In the study of Zincir &Kaya (2009), there was a significant correlation between the nurses' educational status and their knowledge scores ($p < 0.05$); it was stated that the nurses having a bachelor or postgraduate degree known patient rights better than the nurses having a high school diploma and an associate's degree. Although it was determined that there were a study (Zincir &Kaya, 2009) having similar results and a study having different results (Kavak et al., 2014), the results showed that the nurses with a bachelor's degree had higher sensitivity to patient rights. In addition, the sensitivity of the nurses having a bachelor degree to patients' right of 'approval' was higher so we suggest that the nurses having a bachelor degree adopt and take care this right more.

Table 5. Distribution of the scores of the nurses on the overall BRPF scale and its subscales according to their education status

BRPF Subscales			**Education Status**				**KW H Test**	
		n	Mean	SD	Med	Min - Max	H	p
Human Rights and Values in Health Services	HS/VHS graduate	16	4,69	1,4	5	3 - 6	3,73	0,155
	Associate's Degree	28	4,39	1,42	4	3 - 7		
	Bachelor Degree	10	3,6	1,07	3	3 - 6		
Informing	HS/VHS graduate	16	5,06	1,61	5,5	3 - 8	0,57	0,753
	Associate's Degree	28	5,0	1,44	5	3 - 8		
	Bachelor Degree	10	4,6	1,43	5	3 - 6		
Consent	HS/VHS graduate	16	5,56	1,59	6	3 - 8	8,99	0,011*
	Associate's Degree	28	6,5	1,4	6	3 - 9		

Table 5. (Continued)

BRPF Subscales		**Education Status**					**KW H Test**	
		n	Mean	SD	Med	Min - Max	H	p
	Bachelor Degree	10	4,7	2	4	3 - 9		
Privacy and Private Life	HS/VHS graduate	16	4,5	1,26	4,5	3 - 6	2,31	0,315
	Associate's Degree	28	4,54	1,17	4,5	3 - 6		
	Bachelor Degree	10	3,9	1,29	3	3 - 6		
Care and Treatment	HS/VHS graduate	16	6,63	1,59	6	4 - 11	3,84	0,146
	Associate's Degree	28	6,14	1,38	6	3 - 9		
	Bachelor Degree	10	5,3	1,57	5,5	3 - 8		
Consent	HS/VHS graduate	16	7,13	2,55	6,5	3 - 12	0,51	0,775
	HS/VHS graduate	28	7,25	1,67	7,5	3 - 11		
	Associate's Degree	10	6,9	1,73	7	5 - 10		
Total	Bachelor Degree	16	33,56	8,09	33,5	19 - 47	4,12	0,127
	HS/VHS graduate	28	33,82	5,21	34	24 - 42		
	Associate's Degree	10	29,0	7,12	25	22 - 40		

*p < 0,05; HS: High School; VHS: Vocational High School; Med.: Median; Min.: Minimum; Max.: Maximum; SD: Standard Deviation;KW: Kruskal Wallis.

There was no statistically significant difference between the mean scores of the nurses on the overall BRPF Scale and it's all subscales according to their working time, the nurses having < 5 working years had a lower mean overall BRPF score; their sensitivity to patient rights was higher. Our results are similar to the results of the study of Savaskan (2006) in which there was no significant difference between the nurses' beliefs about patient rights according to their working years. Savaskan, (2006) stated that the nurses having < 5 working years had a lower mean overall BRPF score. In conclusion, it can be suggested that more place has been given to patient right in the curricula in the recent time because patient rights has been a topical issue, this condition may increase nurses' sensitivity to patient rights; increasing working years cause a decrease in the willingness of nurses to follow innovations.

Table 6. Distribution of the scores of the nurses on the overall BRPF scale and its subscales according to their working years

BRPF Subscales		Working Years					KW H Test	
		n	Mean	SD	Med	Min - Max	H	p
Human rights and values in health services	0-5 years	16	4,0	1,32	3	3 - 6	3,985	0,263
	6-10 years	13	4,38	1,33	4	3 - 6		
	11-15 years	16	4,19	1,42	3	3 - 6		
	≥16 years	9	5,11	1,45	6	3 - 7		
Informing	0-5 years	16	4,63	1,5	5	3 - 8	4,152	0,245
	6-10 years	13	4,54	1,51	4	3 - 7		
	11-15 years	16	5,38	1,54	6	3 - 8		
	≥16 years	9	5,33	1,12	6	3 - 6		
Consent	0-5 years	16	5,06	1,61	5	3 - 8	5,736	0,125
	6-10 years	13	6,54	1,85	6	3 - 9		
	11-15 years	16	6,06	1,61	6	3 - 9		
	≥16 years	9	6,11	1,45	6	3 - 8		
Privacy and private life	0-5 years	16	3,88	1,15	3	3 - 6	7,6	0,055
	6-10 years	13	4,15	1,21	4	3 - 6		
	11-15 years	16	4,81	1,11	5	3 - 6		
	≥16 years	9	5,0	1,22	5	3 - 6		
Care and treatment	0-5 years	16	6,19	1,87	6	3 - 11	1,733	0,63
	6-10 years	13	5,69	1,89	6	3 - 9		
	11-15 years	16	6,44	1,15	6	5 - 8		
	≥16 years	9	6,11	0,6	6	5 - 7		
Application	0-5 years	16	6,69	1,89	6,5	4 - 10	4,78	0,189
	6-10 years	13	6,69	2,32	7	3 - 10		
	11-15 years	16	8,06	1,81	8	6 - 12		
	≥16 years	9	7,0	1,32	7	5 - 9		
Total	0-5 years	16	30,44	6,97	28	23 - 47	4,981	0,173
	6-10 years	13	32,0	7,98	32	19 - 44		
	11-15 years	16	34,94	5,26	36	25 - 42		
	≥16 years	9	34,67	5,52	37	22 - 40		

*p < 0,05; Med.: Median; Min.: Minimum; Max.: Maximum; SD: Standard Deviation; KW: Kruskal Wallis.

Table 7. Distribution of the scores of the nurses on the overall BRPF scale and its subscales according to their education status about patient rights

BRPF Subscales		Education Status about Patient Rights					MW U Test	
		n	Mean	SD	Med	Min - Max	z	p
Human rights and values in health services	Yes	42	4,31	1,37	4	3 - 7	-0,14	0,885
	No	12	4,42	1,51	4	3 - 6		
Informing	Yes	42	4,76	1,38	5	3 - 8	-1,57	0,116
	No	12	5,58	1,68	6	3 - 8		
Consent	Yes	42	5,67	1,73	6	3 - 9	-1,79	0,073
	No	12	6,67	1,37	6	4 - 9		
Privacy and private life	Yes	42	4,48	1,23	5	3 - 6	-0,67	0,503
	No	12	4,17	1,19	4	3 - 6		
Care and treatment	Yes	42	6,02	1,39	6	3 - 9	-0,78	0,438
	No	12	6,5	1,93	7	4 - 11		
Application	Yes	42	6,88	1,93	7	3 - 12	-1,82	0,068
	No	12	8,08	1,78	8	6 - 11		
Total	Yes	42	32,12	6,71	33	19 - 44	-1,23	0,219
	No	12	35,42	6,08	35,5	24 - 47		

*p < 0,05; Med.: Median; Min.: Minimum; Max.: Maximum; SD: Standard Deviation; MW: Mann Whitney.

There was no statistically significant difference between the mean scores of the nurses on the overall BRPF scale and its subscales according to their status for patient rights education (p > 0.05, Table 7). The nurses who got education about patient rights, had lower scores on the subscales of 'human rights and values in health services', 'informing', 'consent', 'care and treatment' and 'application', the nurses who got education were generally more sensitive to patient rights. Teke et al., (2007) showed that there was no statistically significant correlation between the nurses' education status about patient rights and their mean knowledge and attitude scores related to patient rights (p > 0.05). It can be suggested that nurses who are educated about patient rights are more sensitive.

Conclusion

The sensitivity of the nurses in surgical services to patient rights was high; it was determined that the nurses who were at the ages of 29 or younger, undergraduates, worked 1-5 years and got educated about patient rights, had higher levels of patient rights sensitivity. According to these results, more time should be allocated for this subject in the curricula before the graduation and training programs should be planned for nurses after graduation according to their age groups, working years, their status for getting the education on patient rights, for raising the awareness of nurses about the sensitivity to patient rights, for knowing the problems related with patient rights, for providing solution. The effect of provided education on the knowledge, attitudes and behaviors of nurses should be monitored. It may also be beneficial to conduct further studies in larger groups, including other members of the healthcare team, for determining the different factors that can affect the sensitivities of surgical nurses to patient rights.

References

Acaroglu, R. (2014). Reliability and Validity of Turkish Version of the Nurses Professional Values Scale – Revised. *Journal of Istanbul University Florence Nightingale School of Nursing, 22*, 8-16.

Ardahan, M. (2003). Advocating role of nurses. *Journal of Cumhuriyet University School of Nursing, 7*, 23-7.

Carazo, S. (2000). Ethics in perioperative nursing. *Journal of Cumhuriyet University School of Nursing, 4*, 1-15.

Erbil, N. (2009). Developing scale of patient' rights using attitude. *International Journal of Human Sciences, 6*, 825-37.

Erzincanlı, S. & Zaybak, A. (2015). An examination of patient's attitudes toward the utilization of patient rights. *Journal of Ego University Nursing Faculty, 31*, 39-51.

Kavak, M., Coşunçelebi, Ş., Taşova-Değerli, M., Bozkuş, A. (2014). Assessing nurses level of knowledge in regards to patient rights and a sample practice. *Health Care Academician Journal, 1*, 141-45.

Khalaf, S.K., Al-Asadi, J.N., Adeb, A.H., Shami, S.A., Al-Shamry, H. (2014). Knowledge and attitudes towards patient's rights among health care providers in prımary care health centers in Basrah. *International Journal of Medicine and Pharmaceutical Sciences, 4*, 7-14.

Ozakar Akca, S., Akpinar, Y.Y., Habbani, T. (2015). Knowledge and attitudes of nurses regarding patient rights: A Corum/Turkey sample. *Rev Assoc Med Bras., 61*, 452-57.

Ozer, N., Karaman-Ozlu, Z., Sarutaş, S. (2009). Do the surgical clinic inpatients know their rights? *Journal of Ataturk University School of Nursing, 12*, 19-28.

Oztaş, B. & Iyigün, E. (2016). Determination of the attitudes to use the patient's rights of patients undergone cardiovascular surgery. *Gulhane Medical Journal, 58*, 256-60.

Patients' Rights Regulation, *Official Gazette* Date: 01.08.1998, Official Gazette Number: 23420 Accessed: 11.08.2018.

http://www.saglik.gov.tr/TR,10461/hasta-haklari-yonetmeligi.html.

Savaşkan, F. (2006). *Determination of sensitivity on patient rights of nurses and physicians working at public hospitals in TRNC*. İstanbul University, Institute of Health Sciences, Nursing Basics. Master Program. Istanbul.

Tanrıverdi, H. & Ozmen, M.E. (2011). Effect of patient rights knowledge level on hospital service quality. *TSA, 15*, 85-109.

Teke, A., Uçar, M., Demir, C., Çelen, Ö., Karaalp, T. (2007). Evaluation of knowledge and attitudes of the nurses working in a training hospital about patients' rights. *TAF Preventive Medicine Bulletin, 6*, 259-66.

Temiz, Z., Ozturk, D., Unver, S., Tohumat, Ş.G., Akyolcu, N., Kanan, N., Nur, F. (2017). Determining ethical sensitivity of nurses employed in surgical units. *Journal of Anatolia Nursing and Health Sciences, 20*, 83-9.

Tümer, A.R., Karacaoğlu, E., Akçan, R. (2011). Cerrahide aydınlatılmış onam ile ilgili sorunlar ve çözüm önerileri. [Problems related to informed consent and solutions in surgery.] *Turkish Journal of Surgery, 27*, 191-97.

Zaybak, A., Eşer, İ., İsmailoğlu, E.G. (2012). An analysis of the attitude of patients towards using their patient rights at a university hospital. *Journal of Istanbul University Florence Nightingale School of Nursing, 20,* 104-11.

Zincir, H. & Kaya-Erten, Z. (2009). Knowledge levels of the health care employees working at the primary health institutions about patient rights. *International Journal of Human Sciences, 6,* 877-84.

In: Advances in Health and Natural Sciences ISBN: 978-1-53614-639-4
Editors: Burcu Yuksel et al.

Chapter 14

CONTROL AND ERADICATION OF ANIMAL BRUCELLOSIS

Mustafa Sencer Karagul*
Kartepe Vocational School of Equine Science,
Kocaeli University, Kocaeli, Turkey

ABSTRACT

Brucella spp. causes brucellosis, which is one of the most common zoonotic diseases and which brings about important problems related to health and economy. Direct or indirect contact with infected animals or their products is the main way of transmission of the infection. If brucellosis in animal reservoirs can be controlled, it will be easier to see a decline in the incidence of human brucellosis as well. The primary means of fight against animal brucellosis includes test-and-slaughter programs, routine screening, and control of animal movement, hygiene, herd management, and vaccination. In many countries, achieved success in control and eradication of brucellosis has a governmental regulation/legislation background. Control and eradication require effective collaboration between veterinarians and all the sectors of public health. The aim of this chapter is to reveal general aspects of control and

* Corresponding Author Email: msencerk@hotmail.com.

eradication of this disease and it specifically gives information about the case in Turkey.

Keywords: Brucellosis, control, eradication, vaccine, vaccination

INTRODUCTION

Almost 2/3 of human pathogens are zoonotic and more strikingly, ¾ of emerging and re-emerging diseases in humans are also zoonoses (Godfroid et al., 2013). Some precautions against specific zoonoses and zoonotic agents in animals and animal products used to prevent outbreaks of food-borne infections and intoxications have been revealed by the European Union (EU) through some specific regulations (Council Directive 92/117/EEC). Problems related to brucellosis clearly illustrate the need for a 'One World, One Health' strategy because if the public and animal health authorities work together, control and subsequent eradication of the disease could be achieved (Perez-Sancho et al., 2015). The term 'eradication,' in veterinary medicine, means eliminating an infectious agent and up to now very few diseases have been eradicated. On the other hand, 'control of a zoonotic disease,' refers to all the measures taken in order to decrease the incidence and prevalence of the disease in an animal population (Minas, 2006).

The control of animal brucellosis has been successfully implemented with the help of the combination of vaccination and test-and-slaughter programs in the developed countries. They also succeeded in controlling human brucellosis via milk pasteurization in combination with effective surveillance of the disease and the control of animal movement. (McDermott &Arimi, 2002; Pappas et al., 2006). For this reason, it has been recommended that communities' needs and perceptions should be included in any disease control strategies (Marcotty et al., 2009). For any strategy used for controlling and eradicating brucellosis to be effective, it is necessary for farmers to support the process and collaborate with other stakeholders (Ragan, 2002). It is also necessary for the sectors of public health and veterinarians to collaborate (Yumuk & O'Callaghan 2012).

CONTROL AND ERADICATION

Decreasing the prevalence and incidence will lead to minimal effects on human health and regional economy. For brucellosis, the control measures taken should include monitoring brucellosis-free flocks and preventing the spread of the infection, reducing the rate of infection, and making animals more resistant (Minas, 2006). According to OIE, there are some specific requirements for a country/zone to be considered as free from brucellosis (Resmi Gazete, 2009; Perez-Sancho et al., 2015):

- Bovine brucellosis should be declared as notifiable
- An official veterinary control system including the whole bovine population should be established
- The flock prevalence of the disease should be lower than 0.2%
- There should not be any vaccination that has been conducted in the last three years
- All flocks should be subject to periodical serological testing
- All reactors should be culled
- New animals should be introduced to the regions which are parts of officially brucellosis-free herds

In order for an eradication program of brucellosis to be implemented, flocks should be under strict surveillance and movement control, individual identification of animals should be supplied and veterinary service, which is effective and well organized, should be adopted for surveillance and laboratory testing in place (Minas, 2006). When choosing the best approach to deal with the disease, the prevalence of animal brucellosis should not be considered as the only parameter by the decision makers because in order to use the most sufficient control strategy in each region, other epidemiological and economic aspects should also be taken into consideration (Perez-Sancho et al., 2015).

Even if all the tools to be used to eradicate the disease are existent in eradication and control program, adequate surveillance is a critical first step (Ragan, 2002; Yumuk & O'Callaghan, 2012). Permanent identification of animals is a fundamental element in a successful surveillance program (Ragan, 2002). After the initial phase, where the infection is controlled through obligatory vaccination, vaccination is restrained step by step and it is eventually banned; however, in order to eradicate the infection, a 'test-and-slaughter' policy is carried out (Godfroid et al., 2013).

The most influential strategies used to control this pathology in livestock include strict biosafety, management precautions and test-and-slaughter strategy. There are 3 main strategies which are used in brucellosis control of domestic animals: hygienic measures, test-and-slaughter programs, immunization. As optimal results are gathered through using at least two of them together, the use of only one of these measures is not effective enough (Nicoletti, 2010). To be able to be successful in the program, which includes animal identification, animal movement control, economic compensations, and other complimentary tools should also be added to these already-known strategies (Crespo et al., 2012).

In terms of biosafety and management, what is necessary includes both finding the disease and eliminating it and also doing it before it reaches susceptible herds. If there is success in identification, control and eradication of the disease before it spreads, then it is possible for eradication to be achieved. Understanding of the epidemiology and pathogenesis of brucellosis in eradication is a very important step (Ragan, 2002). Precautions of hygiene and management against brucellosis must emphasize decreasing the possibility of contact with viable *brucella* which includes animals that are infected and contaminated environments. For this reason, in order to prevent the disease from entering in an epidemiological unit, suitable biosecurity precautions should be taken. Hygiene precautions are used to decrease the possibility of contact with viable bacteria and they are also used to control and constrain the bacterial load in the environment. (Perez-Sancho et al., 2015).

It is believed that the existence of the disease in wild animals depends on its existence in domestic animals. The point of connection is that fact that aborted foetusses and foetal membranes are carried by carnivores and it also causes the disease to spread to domestic herds and pastures which are free from brucellosis (Thimm, 1982). Only if infected animals' prevalence is 2% or less, the flocks are kept in a strictly controlled environment and they are protected from the re-entry of infection is it possible for an eradication program including a test-and-slaughter policy to be implemented on the related economic grounds (Minas 2006, Yumuk & O'Callaghan, 2012).

The main objective of the test-and-slaughter approach is detecting and removing possible sources of infection, which will automatically prevent the circulation of *brucella* strains. Because of the silent carriers that are keeping the pathogen in the flock, there might be abortion storms if the immunity of the herd decreases. This strategy seems most useful when there are necessary economic resources and veterinary expertise in places where the prevalence of the disease is low (Perez-Sancho et al., 2015).

Despite the fact that the counter actions against brucellosis in small ruminants have been characterized well, socio-economic factors affect the selection of the most appropriate control measures in most endemic areas. For instance, because pastoral ecosystems, which are traditional ways of management used in low-income areas in which *B. melitensis* is endemic, have led to the continuation of the disease as it is difficult to detect the disease in the early stages and to take necessary control precautions (Perez-Sancho et al., 2015). It was stated that it is possible to implement eradication through a test-and-slaughter policy if sheep and goats are controlled strictly and the type of husbandry is not extensive and grazing of herds is not common (Minas, 2006).

It must also be taken into consideration that the tests used in diagnosis cannot detect all infected animals and they might give false negative results because of the period of incubation, latency or the criteria that are considered for the interpretation of the results (Minas, 2006). For this reason, for the success of the test-control strategies used in the eradication programs, the certainty of the approval of laboratories and diagnostic tests are mentioned as the important factors (Perez-Sancho et al., 2015; Ragan, 2002). For instance, it is stated that the silent carriers of the disease, which are the infants infected congenitally in utero or through the ingestion of contaminated milk or colostrum, may still be seronegative by the time a problem in their reproductivity appears (Perez-Sancho et al., 2015). Therefore, it is necessary to integrate laboratory testing with epidemiological investigation in order to overcome false-positive findings of laboratories as *Brucella* bacteria might be used as potential bioterror agents (Godfroid et al., 2005).

In the case of bioterrorism, it is very important to differentiate between true brucellosis and other diseases which appear in the differential diagnosis of brucellosis. Among these are infections which appear because of cross-reactive bacteria in the serology of brucellosis, such as YO9 (Gourdon et al., 1999). It was stated that there might be the possibility of having laboratory-acquired infections due to the inhalation of infective aerosols which are produced accidentally by microbiological techniques. Thus, it is necessary to utilize biosafety level 3 cabinets in the laboratories which include work related to *Brucella* spp. (Godfroid et al., 2005).

At present, trading takes place between countries which are 'officially free from brucellosis'. Alternatively, the herds/flocks, which are brucellosis-free, might also be preferred for trading. But, there are also some other possibilities; namely, traded animals might be recently infected but the infection might not have been detected yet, they might be infected during transportation or when they are together with other animals at market places. It shows that tests which are recommended for trading can detect acute infections in animals. It must also be taken into account that the only diagnosis of certainty is bacteriology (Godfroid et al., 2013).

Generally, endemic regions are defined as the ones that are structurally weak and the ones which might become worse as proficient veterinary service requires the use of current vaccines effectively. This requirement relates in part to some of the limitations of currently available brucellosis vaccines, and it seems likely that a perfect vaccine could greatly facilitate control and eradication (Godfroid et al., 2011). It was stated that if the population's resistance to disease could be increased with the help of vaccination, it could be possible to control brucellosis. For this reason, vaccination is regarded as more acceptable and efficient when compared to other methods used for this purpose. Moreover, depending upon the vaccines' practical use to eliminate the clinical signs of brucellosis, the environment would be less contaminated and the population's risk to the infection's agent would decrease (Minas, 2006).

TREATMENT

Winning the fight against human brucellosis depends upon the control and eradication of animal brucellosis (Boschiroli et al., 2001; Corbel, 1997; Godfroid et al., 2005). Control and prevention are significant for animals; when it comes to humans, treatment gains more importance. Brucellosis is mentioned as 'cattle brucellosis' and 'sheep-goat brucellosis' in notifiable animal diseases list of 'the Regulation of Reporting of Notifiable Animal Diseases' (Resmi Gazete, 2011). As stated in the *Regulation of Fighting Brucellosis*, it is obligatory to notify the authorities in the case of the existence or the suspicion of brucellosis in cattle, sheep and goats. In the same regulation, it is also pointed out that it is prohibited to take action to treat brucellosis in cattle, sheep and goats (Resmi Gazete, 2009).

Antibiotics have been utilized with success in the treatment of some dogs affected by *B. canis*. Even though it is required to use two different antibiotics simultaneously, the use of only enrofloxacin seemed to be successful in one case. It has also been stated that no treatment has proved to be successful to eliminate *B. canis*. The organism of the disease might appear to have disappeared but it is possible for it to reappear in tissues such as lymph nodes, spleen, uterus and prostate (CFSPH, 2012). In a study by Wanke et al. (2006), enrofloxacin could not effectively treat canine brucellosis but it could help them remain fertile and it also prevented them from aborting and infecting the puppies and spreading the microorganisms during parturition.

When it comes to humans, as *Brucella* bacteria can adapt to the environmental conditions that are met during their replicative niche (Seleem et al., 2008) and their intracellular localization, the rates of treatment failure and relapse are high and they are changeable depending on the combination of drugs and the compliance of patients (Seleem et al., 2010). Because it is highly probable for monotherapies with single antibiotics to relapse, the best treatment for brucellosis is using a combination of two antibiotics. The best option, especially for the acute brucellosis cases, seems to be utilizing doxycycline in combination with streptomycin (DS) as it leads to fewer side effects and relapses (Matthew & Leland, 2004; Seleem et al., 2010). World Health Organization (WHO) has regarded DS combination as the standard treatment against brucellosis for years. In 1986, joint FAO/WHO committee suggested using doxycycline and rifampin (DR) together for 6 weeks for the treatment of adult acute brucellosis (WHO, 1986). However, the studies comparing and contrasting the effects of DR with traditional DS found out that DR, particularly in patients suffering from acute brucellosis, is not as effective as DS (Solera et al., 1997).

If the treatment is inadequate, severe and debilitating chronic courses and long-term sequelae may appear. For this reason, the most important aim is two-folded: controlling acute illness and preventing some possible complications and relapses. In the treatment, utilizing at least two synergistic antibiotics, including doxycycline, rifampin, streptomycin (or other aminoglycosides) or trimethoprim-sulfamethoxazol is a necessity for prolonged chemotherapeutic regimens (Godfroid et al., 2011). After the treatment is over, titers will decrease step by step but IgG antibodies decrease faster than IgM antibodies. Even though there is no active infection, in some cases, low tiers of IgM antibodies might persist for months or years (Doganay & Aygen, 2003; Matthew & Leland, 2004). For this reason, measuring IgG is traditionally used to see whether the therapy is effective or not (Matthew & Leland, 2004).

VACCINATION

Immunization against agents of infection has become a key element in the process of controlling the most significant infectious diseases since Eduard Jenner's description of first vaccine against smallpox in 1796 (Perez-Sancho et al., 2015). The only way to control and eradicate brucellosis is believed to be vaccinating all the hosts which are susceptible and eliminating all of the infected animals in the regions in which the prevalence of the disease is high (Briones et al., 2001).

The most common vaccines which are used against bovine brucellosis are *B. abortus* strain 19 and RB51, which were recently approved by USDA and which do not interfere with serological diagnoses (Moriyon et al., 2004). As can be seen in all highly contagious diseases, the effect of vaccination augments as the coverage of vaccination increases. Only if the vaccine is a high quality one and if it can be administered to at least 80% of the risk-group animals, sufficient protection can be possible (Minas, 2006).

B. abortus S19 and *B. melitensis* Rev1, which are considered to be the best two attenuated vaccines developed in the past century, have both a good degree of residual virulence. Rev1 vaccine is not only used against *B. melitensis* in sheep and goats but also against *B.ovis.* S19 strain has appeared as a result of accidental attenuation, which was due to the long laboratory storage under insufficient conditions and Rev1 strain is a revertant of a streptomycin-dependent mutant with a known ribosomal mutation and possibly other genetic defects (Godfroid et al., 2011). The first step for the eradication of *B. melitensis* infection in the areas, where it is endemic or where the prevelance of the disease is high, is controlling the disease through the vaccination of animals. For small ruminants, the vaccine used is Rev-1 vaccine. An alternative method for brucellosis control, particularly in the endemic areas where the prevelance of the disease is more than 5%, is vaccination of all animals in the flock/herd regardless of their age. There are good results in different areas with sufficient financial resources (Minas, 2006).

There is evidence that Rev1 vaccine is resistant to streptomycin, which is used in the treatment of human brucellosis and S19 vaccine is resistant to penicillin G. In addition, it is stated that RB-51, which was prepared for cattle, has resistance to rifampicin, which is again used against human brucellosis (Grillo et al., 2000). S19 and Rev1 vaccines carry a smooth (S) lipopolysaccharide (S-LPS) and an O-polysaccharide (PS), which is similar to brucellae that are of wild type. In the controlled experiments, only one dose of Rev1 provides 80-100% protection against challenge strains (Barrio et al., 2009).

Rev1 strain is regarded as the reference vaccine to immunize the sheep and goats which are at risk for infection and it needs to be compared and contrasted with other vaccines (OIE, 2009a). The disadvantages of this vaccine include its causing abortion during pregnancy, its interference in serological diagnosis, its virulence for human beings, its resistance to streptomycin and its tendency to dissociate into rough mutants which are ineffective (Godfroid et al., 2011; Seleem et al., 2010). Moreover, because it might be difficult to discriminate the antibody response to vaccination from the ones seen after field infection, they might interfere with control programs. There are still attempts to develop new live attenuated rough vaccines for *B. melitensis,* which are deprived of O-side chain. These vaccines need further evaluation in field experiments (Adone et al., 2008; Blasco, 1997; Garin-Bastuji et al., 1998).

It is possible to partially go beyond these limitations through vaccinating animals conjunctivally when they are younger than 4 months old. It will decrease the amount of interference in serological diagnosis and abortions caused by vaccines. Thanks to a minimal amount of personal protection, Rev1 vaccine becomes safe and there are well-established protocols of quality control (Godfroid et al., 2011). Despite the insufficiency of related studies, it is pointed out that *B. melitensis* in cattle might be controlled with the help of S19 vaccine. All the successful programs in cattle have utilized S19 except for a number of countries with good conditions of geography and management (Godfroid et al., 2011). Thanks to adult vaccination with S19, it is possible to increase herd resistance and to decrease shedding of *B.abortus* at calving or abortion

(Ragan, 2002). When animals are younger than 4 months old, conjunctival vaccination including smaller doses will reduce serological interference and udder infections. S19 vaccine used in cattle is both economical and it is not resistant to antibiotics. Even though it is not as virulent as Rev1 for humans, it needs minimum level of individual protection (Godfroid et al., 2011).

Vaccination of both rams and ewes is the most feasible way of providing the control of *B. ovis* in places having a high prevalence of infection because both of them play an important role in the transmission of infection. There is not a certain vaccine against *B. ovis* but live *B. melitensis* strain Rev1 vaccine is appropriate for stimulating immunity against *B. ovis*. It is also possible to use a single dose of Rev1 (10^9cfu) given subcutaneously or conjunctivally to overcome *B. melitensis* infection (OIE 2015). There is not an existing vaccine for pigs (Godfroid et al., 2011). In China, *B. suis* biovar-2 vaccine is used to immunize pigs but in order to recommend strain 2 vaccine for general use, the results obtained in China should be confirmed (OIE, 2009b).

It is stated that RB51 vaccine might lead to abortions and shedding. Its use should be limited with animals which are not pregnant. RB51 can induce abortions and can be excreted (Moriyon et al., 2004). It is also claimed that RB51 vaccine is more attenuated and as a result, it is considered to be less dangerous than S19 and there are only a few human cases mentioned (Ashford et al., 2004; Godfroid et al., 2011). But for human brucellosis, the number of unreported human cases might be more as the lack of anti-O-PS antibodies makes RB51 difficult to detect in standard laboratory tests. For this reason, it is necessary to keep biosafety practices while utilizing RB51. A further disadvantage of RB51 is the resistance to rifampin since this antibiotic is more preferable than streptomycin or gentamicin for treatment (Godfroid et al., 2011).

There have been several attempts to develop a killed Brucella vaccine for human use but so far, they have not been utilized extensively because there is no long-lasting immunity (Matthew & Leland, 2004). There is not an existing vaccine for human brucellosis (Godfroid et al., 2005; Godfroid et al., 2011; Yumuk & O'Callaghan, 2012). Under the WHO global program on vaccination and immunization, priority should be given to the development of such a vaccine (Godfroid et al., 2005).

THE FIGHT AGAINST BRUCELLOSIS IN TURKEY

In Turkey, the first brucellosis case was identified by Husamettin Kural and Mahmut Sabit Akalin in 1915 in a soldier, who was being treated in Istanbul Kuleli Military Hospital (Celebi, 2003; Cevik, 2001; Golem, 1949; Iyisan et al., 2000). In the Turkish context, the first documented case of brucellosis, which was confirmed by a laboratory was in 1915 despite the fact that it is believed that Florence Nightingale got this disease during the Crimean War in Turkey (Young, 1995; Yumuk & O'Callaghan, 2012). In many regions, pasteurizing fresh milk protects the city population from this disease; however, there is still some risk for people who are in direct contact with infected animals and who consume fresh milk (Kuplulu & Sarimehmetoglu, 2004; Yumuk & O'Callaghan, 2012).

It is stated that the control of the disease is the responsibility of veterinarians as almost all cases of human brucellosis are originated from animals (Nicoletti, 2002; Yumuk & O'Callaghan, 2012). In this context, the precautions against this anthropozoonosis should include the control and eradication of the disease in the animal reservoir (Thimm, 1982). The initial work on the control and eradication of brucellosis started in 1930 for *Brucella abortus* in cattle. In this sense, the Brucella Laboratory established within Ankara Etlik Institute of Veterinary Bacteriology and Serology moved to the Institute of İstanbul Pendik Veterinary Bacteriology and Serology in 1951. *Brucella melitensis* laboratory was founded in 1965 and later on, the production of *B. melitensis* Rev-1 started in 1969 (GTHB, 2012).

In Turkey, 'National Brucella Control and Eradication Project' was implemented in 1984 by the Ministry of Agriculture and Forestry. For this project, a 26-year-period (1984-2010) was determined and vaccination of calves, labs and ewes was the main aim. (GTHB, 2012; Iyisan 2006). After this project, the authorities tried to find out about the prevalence of the disease in cattle and sheep in 2011 and it was revealed that herd prevalence of brucella in cattle is 7.8% and in sheep it is 22.5% (GTHB, 2012).

Experts in the Ministry evaluated the results of the work carried out in 2011 on the prevalence of Brucellosis and focused on the strategies adopted by the countries to eradicate this disease. They decided that mass vaccination is the most feasible method and they also decided that mass vaccination should be through conjunctival vaccination of animals of all ages. In this context, 'The Project of the Control and Eradication of Brucella through Conjunctival Vaccines' started in 2012 and it is planned to continue for 10 years in cattle, for 6 years in sheep and goats (GTHB, 2012). After this mass vaccination which started in 2012, Brucellosis outbreaks and human cases were evaluated. Except for the vaccine-related abort cases in 2013, there are dramatic decreases in outbrakes and very significant declines in human cases. It is possible to state that the continuation of mass vaccination as well as other control and sanitary measures should be the main approach (Saytekin et al., 2015).

If brucellosis incidence in animal reservoirs is controlled, there can be a related decrease in the incidence of human brucellosis as well. After the long and successful efforts of developed countries which were able to control bovine brucellosis, there was also a visible decrease in the incidence of human infection. Therefore, the eradication of brucellosis can be achieved only if all the parties are involved in the process of finding a solution (Seleem et al., 2010).

References

Adone, R., Francia, M., Ciuchini, F. (2008). Evaluation of Brucella melitensis B115 as rough-phenotype vaccine against B. melitensis and B. ovis infections. *Vaccine, 26*, 4913–4917.

Ashford, D. A., di Pietra, J., Lingappa, J., Woods, C., Noll, H., Neville, B., Weyant, R., Bragg, S. L., Spiegel, R. A., Tappero, J., Perkins, B. A.

(2004). Adverse events in humans associated with accidental exposure to the livestock brucellosis vaccine RB51. *Vaccine, 22*, 3435–3439.

Barrio, M. B., Grillo, M. J., Munoz, P. M., Jacques, I., Gonzalez, D., De Miguel, M. J., Marin, C. M., Barberan, M., Letesson, J. J., Gorvel, J. P., Moriyon, I., Blasco, J. M., Zygmunt, M. S., (2009). Rough mutants defective in core and O-polysaccharide synthesis and export induce antibodies reacting in an indirect ELISA with smooth lipopolysaccharide and are less effective than Rev 1 vaccine against Brucella melitensis infection of sheep. *Vaccine*, *27*, 1741–1749.

Blasco, J. M., (1997). A review of the use of B. melitensis Rev 1 vaccine in adult sheep and goats. *Prev. Vet. Med., 31*, 275–283.

Boschiroli, M. L., Foulongne, V., O'Callaghan, D. (2001). Brucellosis: a worldwide zoonosis. *Curr Opin Microbiol, 4*, 58-64.

Briones, G., Inon de Iannino, N., Roset, M., Vigliocco, A., Paulo, P. S., Ugalde, R. A. (2001). Brucella abortus cyclic beta-1, 2-glucan mutants have reduced virulence in mice and are defective in intracellular replication in HeLa cells. *Infect Immun*., *69*, 4528–4535.

Celebi, S. (2003). Bruselloz Epdemiyolojisi. *ANKEM Dergisi, 17*(3), 340-343. [Epidemiology of Brucellosis. *ANKEM Journal, 17*(3), 340- 343].

Cevik, A. M. (2001). Bruselloz Epidemiyolojisi. *ANKEM Dergisi*, *15*(3), 568-570. [Epidemiology of Brucellosis. *ANKEM Journal, 15*(3), 568-570].

Corbel, M. J. (1997). Brucellosis: an Overview. *Emerg Infect Dis., 3*(2), 213-221.

Crespo León F, Saez Llorente J. L, Reviriego Gordejo F. J, Rodriguez Ferri E. F., Duran, FM. (2012). Complementary tools for the control and eradication of caprine and ovinebrucellosis in the European Union. *Rev Sci Tech OIE, 31*(3), 985-996.

Garin-Bastuji, B., Blasco, J. M., Grayon, M., Verger, J. M., (1998). Brucella melitensis infection in sheep: present and future. *Vet Res., 29*, 255–274.

Gıda, Tarım ve Hayvancılık Bakanlığı (GTHB)-Gıda ve Kontrol Genel Müdürlüğü. (2012). *Brusellanın Konjuktival Aşı İle Kontrol ve Eradikasyon Projesi-Genelge No: 2012/03*. [Ministry of Food,

Agriculture and Livestock (MFAL) - General Directorate of Food and Control (2012). *Control and Eradication Project of Brucella with Conjunctival Vaccine - Circular No: 2012/03*].

Godfroid J., Cloeckaert A., Liautard J. P., Kohler S., Fretın D., Walravens K., Garin-Bastuji B., Letesson J. J. (2005). From the discovery of the Malta fever's agent to the discovery of a marine mammal reservoir, brusellosis has continously been a re-emerging zonosis. *Vet Res., 36*, 313–326.

Godfroid, J., Scholz, H. C., Barbier, T., Nicolas, C., Wattiau, P., Fretin, D.,Whatmore, A. M., Cloeckaert, A., Blasco, J. M., Moriyon, I., Saegerman, C., Mumak, J. B., Al Dahouk, S., Neubauern, H., J.-J. Letesson, J-J. (2011). Brucellosis at the animal/ecosystem/human interface at the beginning of the 21st century. *Prev Vet Med., 102*, 118-131.

Godfroid, J., Al Dahouk, S., Pappas, G., Roth, F., Matopeg, G., Mumah, J., Marcotty, T., Pfeiffer, D., Skjerve, E.ve ark. (2013). A "One Health" surveillance and control of brucellosis in developin gcountries: Moving away from improvisation. *Comp Immunol Microbiol Infect Dis., 36*, 241– 248.

Golem, S. B. (1949). Brusellozun Memleketimizdeki Durumu. *Türk Hijyen Tecrübi Biyoloji Dergisi*, *9*, 32-63. [The Status of Brucellosis in Our Country. *Turkish Journal of Hygiene, 9,* 32- 63].

Gourdon F., Beytout J., Reynaud A., Romaszko J. P., Perre D., Theodore P., Soubelet H., Sirot J. (1999). Human and animal epidemic of Yersinia enterocolitica O:9, 1989–1997, Auvergne, France, *Emerg Infect Dis.,5,* 719–721.

Grillo MJ, Bosseray N, and Blasco JM. (2000). In vitro markers and biological activity in miceof seed lot strains and commercial Brucella melitensis Rev 1 and Brucella abortus B19vaccines. *Biologicals*, *28*(2), 119-127.

Iyisan, S. (2006). Hayvanlarda Brursellozun Epidemiyolojisi. *Zoonatik Hastalıklar Sempozyumu.* [Epidemiology of Brurselloz in Animals. *Zoonatic Diseases Symposium*].

Iyisan, A. S., Akmaz, O., Gokcen Duzgun, S., Ersoy, Y., Eskiizmirliler, S., Guler, L., Yurtalan, S. (2000). Türkiye'de Sığır ve Koyun Brucellosis'in Seroepidemiyolojisi. *Pendik Veteriner Mikrobiyoloji Dergisi*, *31*(1), 21-75. [Brucellosis in Cattle and Sheep Seroepidemiology in Turkey. *Pendik Journal of Veterinary Microbiology, 31*(1) 21-75].

Kuplulu, O. ve Sarimehmetoglu, B. (2004). Isolation and identification of Brucella spp. in ice cream. *Food Control, 15,* 511-514.

Marcotty, T., Matthys, F., Godfroid, J., Rigouts, L., Ameni, G., van Pittius, N. G., Kazwala, R., Muma, J., van Helden, P., Walravens, K., de Klerk, L. M., Geoghegan, C., Mbotha, D., Otte, M., Amenu, K., bu Samra, N., Botha, C., Ekron, M., Jenkins, A., Jori, F., Kriek, N., McCrindle, C., Michel, A., Morar, D., Roger, F., Thys, E., van den Bossche, P. (2009). Zoonotic tuberculosis and brucellosis in Africa: neglected zoonoses or minor public-health issues? The outcomes of a multi-disciplinary workshop. *Ann Trop Med Parasito.*, *103*, 401–411.

Matthew, L. L., Leland, S. R. (2004). Brucellosis. *Infect Dis Clin Pract*, *12*, 7-14.

McDermott, J. J., Arimi, S. M. (2002). Brucellosis in sub-Saharan Africa: epidemiology, control and impact. *Vet Microbiol*, *90*, 111–134.

Minas, A. (2006). Control and eradication of brucellosis in small ruminants. *Small Rumin Res, 62*, 101-107.

Moriyon, I., Grillo, M. J., Monreal, D., Gonzalez, D., Marin, C., Lopez-Goni, I., Mainar-Jaime, R. C., Moreno, E., Blasco, J. M. (2004). Rough vaccines in animal brucellosis: structural and genetic basis and present status. *Vet Res, 35*, 1–38.

Nicoletti P. (2010). Brucellosis: past, present and future. *Prilozi.*, *31*(1), 21-32.

Pappas, G., Papadimitriou, P., Akritidis, N., Christou, L., Tsianos, E. (2006). The new global map of human brucellosis. *Lancet Inf Dis.*, *6*(2), 91-9.

Perez-Sancho, M., Garcıa-Seco, T., Domınguez, L. ve Alvarez, J. (2015). Control of Animal Brucellosis — The Most Effective Tool To Prevent Human Brucellosis. Retrieved, 09.24.2016, http://www.intech

open.com/: http://www.intechopen.com/books/updates-on-brucellosis/control-of-animal-brucellosis-the-most-effective-tool-to-prevent-human-brucellosis.

Ragan, V. E. (2002). The Animal and Plant Health Inspection Service (APHIS) brucellosis eradication program in the United States. *Vet Microbiol., 90*, 11-18.

Resmi Gazete (2009). Bruselloz ile Mücadele Yönetmeliği., Sayısı: 27189. [*Official Newspaper* (2009). Regulation on Fighting with Brucellosis., Number: 27189].

Resmi Gazete (2011). İhbarı Mecburi Hayvan Hastalıkları ve Bildirimine İlişkin Yönetmelik. Sayısı: 27823. [*Official Newspaper* (2011). Implementing Regulation on Notified Animal Diseases and Notification., Number: 27823].

Seleem, M. N., Boyle, S. M., Sriranganathan, N. (2008). Brucella: A pathogen without classic virulence genes. *Vet Microbiol., 129*, 1-14.

Seleem M. N., Boyle S. M., Sriranganathan N. (2010). Brucellosis: A re-emerging zoonosis. *Vet Microbiol., 140*(3-4), 392-398.

Solera, J., Martinez-Alfaro, E., Espinosa, A., (1997). Recognition and optimum treatment of brucellosis. *Drugs, 53*, 245–256.

The Center for Food Security &Public Health (CFSPH). (2012). *Canine Brusellosis: Brucella canis.* Retreived 11.19.2015, http://www.cfsph.iastate.edu/Factsheets/pdfs/brucellosis_canis.pdf.

Thimm, B. (1982). *Brucellosis- Distribution in Man, Domestic and Wild Animals.* Berlin: Springer.

Wanke, M. M., Delpino, M. V. ve Baldi, P. C. (2006). Use of enrofloxacin in the treatment of canine brucellosis in a dog kennel (clinic trial). *Theriogenology, 66*, 1573-1578.

World Health Organisation for Animal Health (OIE). *Terrestrial Manual* Chapter 2.7.2. (2009a). Caprine and ovine brucellosis. Retrieved, 09.24.2016, http://web.oie.int/eng/normes/MMANUAL/2008/pdf/2.07.02_CAPRINE_OVINE_BRUC.pdf.

World Health Organisation for Animal Health (OIE). *Terrestrial Manual* Chapter 2.8.5. (2009b). Porcine Brucellosis. Retrieved, 09.24.2016 http://web.oie.int/eng/normes/MMANUAL/2008/pdf/2.08.05_PORCINE_BRUC.pdf.

World Health Organisation for Animal Health (OIE). *Terrestrial Manual* Chapter 2.7.9. (2015). Ovine Epididymitis (Brucella ovis) Retrieved, 09.24.2016, http://web.oie.int/eng/normes/MMANUAL/2008/pdf/2.07.09_OVINE_EPID.pdf.

WHO. (1986). *Joint FAO/WHO Expert Committee on Brucellosis.* Erişim, 24.09.2016, http://apps.who.int/iris/bitstream/10665/40202/1/WHO_TRS_740.pdf Geneva: World Health Organization.

Yumuk Z., O'Callaghan D. (2012). Brucellosis in Turkey-an overview. *Int J Infect Dis., 16*(4), 228-35.

Young, D. A. (1995). Florance Nightingale's fewer. *BMJ*, *311*, 1697-1700.

In: Advances in Health and Natural Sciences ISBN: 978-1-53614-639-4
Editors: Burcu Yuksel et al.

Chapter 15

SMALL RUMINANT MORBILLIVIRUS

Ahmet Sait*
Pendik Veterinary Control Institute, Viral Diagnosis Laboratory,
Istanbul, Turkey

ABSTRACT

Small ruminant morbillivirus (SRM), also known as 'goat plague', 'Kata', 'syndrome of stomatitis-pneumoenteritis' or 'ovine rinderpest', is a viral disease of sheep, goats and small wildlife ruminants characterized by fever, sores in the mouth, diarrhea, pneumonia, and sometimes death. The disease is considered to be a major barrier to the development of sustainable agriculture across the developing countries in the world. The global eradication of the disease has been targeted by the World Organisation for Animal Health (OIE) and the Food and Agriculture Organisation (FAO) by 2030. This chapter aims to share current information about the disease.

Keywords: goat, Kata, morbillivirus, peste des petits ruminants virus, sheep

* Corresponding Author Email: saitvet@gmail.com.

INTRODUCTION

The genus *Morbillivirus* belongs to the virus family *Paramyxoviridae*, contains highly infectious viruses that spread via the respiratory route and causes immune suppression. Because of these features, they have a potential causing large outbreaks associated with high morbidity and mortality in previously unexposed humans and animals. *Small ruminant morbillivirus, Rinderpest morbillivirus, Canine morbillivirus, Newcastle Disease Virus* and *Measles morbillivirus* are some of these viruses. Recently, new morbilliviruses, *Phocine morbillivirus* and *Cetacean morbillivirus,* have been detected in marine mammals (Chauhan et al., 2009).

Small ruminant morbillivirus (SRM), also named as *Peste des Petits Ruminants Virus (PPRV)*, causes a severe disease, Peste des petits ruminants (PPR), in sheep, goats and small wildlife ruminants. PPR was first described in the Ivory Coast in West Africa (Gargadennec & Lalanne, 1942). The disease has since been recognized as endemic in West and Central Africa (Scott, 1981) and in the north-east of the continent, Kenya and Uganda (Wamwayi et al., 1995), Sudan (Haroun et al., 2002) and Ethiopia (Reoder et al., 1994; Abraham et al., 2005). In 1987, it appeared in the Middle East and has since then been confirmed in Jordan (Lefevre et al., 1991), Pakistan (Amjad et al., 1996), southern India (Nanda et al., 1996), Turkey (Ozkul et al., 2002; Dagalp et al., 2015; Sevik & Sait, 2015). The detection of the disease in Tunisia (Ayari-Fakhfakh et al., 2010), Morocco (FAO, 2009) and Algeria indicates that the infection has reached the border of the European Union (Banyard et al., 2010). Moreover, the detection of the disease in the Thrace region of Turkey in 1996 has led to questioning the potential spread of the disease to other parts of Europe (Ozkul et al., 2002; Banyard et al., 2010). While this article was being prepared, the *World Organisation* for *Animal Health (*OIE) informed that the disease was detected in Bulgaria for the first time (OIE, 2018).

ETIOLOGY

The causal agent, *Small ruminant morbillivirus* (SRM), *belongs to the genus Morbillivirus in the family Paramyxoviridae* (Gibbs et al., 1979; Bailey et al., 2005). SRM has single stranded nonsegmented negative sense RNA genome, which encodes six structural proteins: nucleocapsid protein (N), phosphoprotein (P), matrix protein (M), fusion (F) protein, hemagglutinin (H), and polymerase (L) protein. The P gene also encodes two non-structural proteins, C and V, which are thought to function in viral evasion of the host immune responses (Nanda & Baron, 2006). Due to its biological and antigenic properties, it has been closely related to *Rinderpest morbillivirus* and other morbilliviruses (*Canine morbillivirus, Measles morbillivirus, Phocine morbillivirus* and *Cetacean morbillivirus*) (Gibbs et al., 1979; Baron et al., 2011). Molecular techniques and nucleic acid sequence analysis studies have proven that the SRM is completely separate from the *Rinderpest morbillivirus* (Gopilo, 2005). As a result of molecular epidemiological studies, the SRM has been divided into 4 different genetic lineages according to the F and N gene partial sequence analysis. It has been observed that there are nucleotide changes between SRM isolates in these 4 different genetic lineages that are compatible with the geographical distribution but do not express a separate serotype or strain (Chard, 2008). According to geographical distribution, lineages I and II have been found in West Africa, whereas lineage III has been found in eastern Africa, Arabian Peninsula and southern India. Lineage IV is found in the Middle East, south Asia and China (Dhar et al., 2002; Banyard et al., 2010; Kwiatek et al., 2011). Used as vaccine strains, Nigeria75/1 strain was located in lineage I, and Sungri96 strain was located in lineage IV (Shaila ve ark., 1996; Dhar ve ark., 2002; Muthuchelvan, 2006).

HOSTS

SRM is an important pathogen of small ruminants and clinical disease is seen in sheep and goats and has been described in wild small ruminants

(Hamdy et al. 1976; Reoder and Obi 1999). However, subclinical infections have been reported in cattle, which formed cross protection against virus (Diop et al. 2005). Buffaloes, camels and pigs may be infected with the virus, but show little or no clinical symptoms (Chauhan et al., 2009). Antibody production against virus was also detected in camels (Roger et al., 2001). SRM could infect wild animals both naturally and experimentally. Antelope and other small wild ruminants including Laristan sheep (Ovis gmelini laristanica), Dorcas-type gazelles (Gazella granti), gemsbok (Oryx gazelle), and the Nubian ibex (Capra ibex nubiana) are severely affected by the disease (Hamdy & Dardiri, 1976; Abu Elzein et. al. 2004; Gur & Albayrak, 2010). Balamurugan et al. (2012) reported that the virus obtained from frozen tissue taken from Asian lion (Panthera leo persica) belongs to lineage IV according to the partial sequnence of N, F and M genes. At the same time, this report was the first report on SRM isolation in Asian lions. The disease does not pose any risk to humans (OIE, 2013).

Transmission

Contagion of SRM from infected to susceptible animals is via by direct contact, close contact and oral routes (Chauhan et al. 2009). Since the virus has lost its viability in the outdoors in a short time, the most important way of transmission is the close contact of the patients in the acute febrile period with the susceptible animals. The aerosol droplets and stools of the infected animals contain high amounts of virus. Small infective droplets spreading to the area, especially by sneezing and coughing, transmit the disease to the susceptible animals through respiration. Besides the close contact, the contaminated bed, feed and water also form an additional source of contamination (Gopilo, 2005). It has been found that viral antigens are scattered in stools until the 11th week after SRM infection (Ezeibe et al., 2008). This suggests that naturally infected animals may pose a significant risk for susceptible populations.

As a current issue, the role of wildlife in the transmission of SRM attracts attention. However, since there is not enough epidemiological data yet, more detailed investigations are needed (Baron et al., 2011).

Clinical Signs

SRM infection is characterized by different clinical patterns ranging from perakut, acute and subacute depending on the various factors such as species and age of the animal, virulence and amount of the virus, secondary infections, etc. (Diallo, 1988; Lefevre ve Diallo, 1990). The mortality rate in the infected herd may vary between 10-90%, and especially in naive population it can be up to 90%. These rates are usually lower in endemic areas, where mortality may be 20% or less, and serosurveillance is sometimes the only indicator of infection. The incubation period varies between 3-9 days depending on the amount of virus and the route of infection (Baron et al., 2011). During the incubation period, the virus replicates in oropharyngeal lymph nodes and then spreads to other tissues and organs. After the incubation period, body temperature increases by 40-41,5°C. A peracute form that results in death a few days after the onset of fever is rare (Gibbs et al., 1979).

In acute form, which is common, the fever lasts for up to a week and the symptoms of the disease are seen in this form. In this period, leucopenia, hyperemia in the mouth mucosa, congestion in the nose and eye mucosa and oculonasal discharge are remarkable. After the fever begins to rise, scales (accumulations of surface skin cells, such as seen in *dandruff*) and erosive lesions begin to appear in the mouth and vulva. Initially, the foci of necrosis in small gray-white color and dot shape grow and spread over time. Then, necrotic areas start to appear on the gums, dental pad, palate, lips, inner aspects of the cheeks and upper surface of the tongue. These areas increase in number and size and join together. The necrotic foci in the mouth and on the cheek rapidly become eroded and covered with yellow-matte white cell destruction. The erosions can spread to the lower face of the tongue and the pharynx. Then, the necrotic areas are covered with exudate released from the inflamed tissue. Excessive salivation and water consumption are seen in sick animals; moreover, these are restless, depressive and anorexic. They have superficial respiration and their breath has a bad smell (Baron et al., 2011). The serous nature of the mouth, nose and eye fluids at the beginning is intense and yellow due to secondary bacterial infections. Diarrhea occurs 1-3 days after the onset of the infection and death generally occurs 5-12 days later. Abortions may occur during infection (FAO, 1999).

The subacute form of the disease is characterized by mucosal discharge that lasts for 2-3 days and does not attract much attention, with a moderate increase in body temperature for 1-2 days. There is no death in this form of the disease (Diallo, 2006).

DIAGNOSIS

The use of rapid, specific, sensitive laboratory methods in the diagnosis of SRM is crucial for the effective implementation of control programs (Diallo et al., 1995). The disease can be confused with cattle plague, foot and mouth disease, pasteurellosis, ectima, sheep and goat pox, pneumoniae, bluetongue, coccidiosis, colibasillosis, poisoning, cryptosporodiosis and pestivirus infections. Although infection is suspected in the light of clinical findings, laboratory tests are essential for differential diagnosis (Diallo, 1988; Scott, 1990). Laboratory tests include viral antigen detection, virus isolation, nucleic acid detection and detection of specific antibodies in serum samples taken at 14 days intervals (Gopilo, 2005). The collection of samples at the correct time is important to achieve diagnosis by virus isolation. Anticoagulant-treated blood and swab samples (conjunctival discharges, nasal secretions, buccal and rectal mucosae) should be taken from live animals in the acute phase of the disease when clinical signs are apparent (OIE, 2013). Tissue samples (pulmonary, spleen, lymph nodes) should be obtained from dead animals for virological diagnosis within the 2 hours after the death. The samples should be delivered to the laboratory in the cold chain as soon as possible.

There are several methods available for the diagnosis of SRM. Some of them are virus isolation, reverse transcription polymerase chain reaction (RT-PCR) and enzyme-linked immunosorbent assay (ELISA). Because it's a cheap method that allows to work with many samples, ELISA is now widely used in the labs (Libeau et al., 1995). Virus isolation in cell culture is a specific technique that needs experience and time, but it has low efficiency. For this reason, it is not widely used in routine diagnosis. RT-PCR allowing to nucleotide sequencing and subsequent phylogenetic analysis and the real time RT-PCR (rRT-PCR) performing quantification of viral loads are the current diagnostic methods for all morbillivirus infections (Forsyth & Barrett, 1995; Shaila *et al.,* 1996; Couacy-Hymann *et al.,* 2002).

PROTECTION AND CONTROL

SRM control measures mainly involve disease surveillance, vaccination, stamping out and to strict control of illegal or uncontrolled animal movements (Abubakar et al., 2011).

Surveillance should aim identification of the prevalance of the disease, new outbreaks and disease free areas, monitoring of the disease trends, and informing what should be made at critical control points. Risk-based surveillance is the most cost-effective way to identify and monitor the disease. While high-risk based surveillance focuses on the detection of incursions at critical points, low risk based surveillance focuses on early detection and the response of disease in areas free from disease (SADC 2012).

Since the prevention of illegal animal movements is difficult to implement successfully in the Middle East, Africa and Asia, where the disease is endemic, vaccination is regarded as the most basic way of controlling PPR (Diallo, 2003). Sheep and goats probably do not have multiple infections with SRM throughout their economic lives. Those animals vaccinated with Nigeria75/1 vaccine strain have up to 100% immunity and 12 months of protection againts the disease (Awa et al.,

2002). Neutralizing antibodies are responsible for immunity. SRM specific neutralizing antibodies were detected on the 16th day of vaccination (Gulyaz et al., 2005). It has been reported that vaccination is safe in field trials and can be used easily even in pregnant animals (Diallo et al., 1995). Maternal antibodies can be detected up to 6 months in lambs and kids born from vaccinated mothers; but it can provide protection for 3.5 months in the lambs and 4.5 months in the kids. For this reason, lambs should be immunized in the 4th month and the kidneys in the 5th month (Awa et al., 2002). It is considered that the immunity is lifelong in animals recovering from the disease (OIE, 2013).

TREATMENT

There is no specific treatment for PPR but antibiotics and other supportive treatment may prevent secondary infections and decrease mortality (OIE 2013).

REFERENCES

Abraham, G. (2005). Antibody seroprevalence against peste des petits ruminants virus in camels, cattle, goats and sheep in Ethiopia. *J. Prev. Vet. Med., 70,* 51-57.

Abubakar, M., Ashiq, S., Zahoor, A. B., Arshed, M. J., Banyard, A. C. (2011). Diagnosis and control strategies for peste des petits ruminants virus: Global and Pakistan perspectives. *Pak Vet J., 31 (4),* 267-274.

Abu Elzein, E. M. E., Housawi, F. M. T., Bashareek, Y., Gameel, A. A., Al-Afaleq, A. I., Anderson, E. C. (2004). Severe PPR infection in Gazelles kept under semi-free range conditionsin Saudi Arabia. *J. Vet. Microbiol. B51 (2),* 68-71.

Amjad, H., Qamar-ul-Islam Forsyth, M., Barrett, T. & Rossiter, P. B. (1996). Peste des petits ruminants in goats. *Pak Vet J., 139,*118-119.

Awa, D. N., Ngagnou, A., Tefiang, E., Yaya, D., Njoya, A. (2002). Post vaccination and colostral peste des petits ruminants antibody dynamics in research flocks of Kirdi goats and Foulbe sheep of North Cameroon. *Prev. Vet. Med. 55,* 265-271.

Ayari-Fakhfakh, E., Ghram, A., Bouattour, A., Larbi, I., Gribaa-Dridi, L., Kwiatek, O., Bouloy, M., Libeau, G., Albina, E. & Cetre-Sossah, C. (2010). First serological investigation of peste-des-petits-ruminants and Rift Valley fever in Tunisia. *Vet J., 187 (3),* 402-404.

Bailey, D., Banyard, A., Dash, P., Ozkul, A., & Barrett, T. (2005). Full genome sequence of peste des petits ruminants virus, a member of the Morbillivirus genus. *Virus Research, 110 (1-2),* 119-124.

Balamurugan, V., Sen, A., Venkatesan, G., Bhanot, V., Yadav, V., Bhanuprakash, V., & Singh, R. K. (2012). Peste des petits ruminants virus detected in tissues from an Asiatic lion (*Panthera leo persica*) belongs to Asian lineage IV. *J Vet Sci, 13 (2),* 203-206.

Banyard, A. C., Parida, S., Batten, C., Oura, C., Kwiatek, O., Libeau, G. (2010). Global distribution of petse des petits ruminants virus and prospects for improved diagnosis and control. *J Gen Virol, 91,* 2885-2897.

Baron, M. D., Parida, S., Oura, C. A. L. (2011). Peste des petits ruminants: a suitable candidate for eradication? *Vet Record, 169*, 16-21.

Chard, L. S., Bailey, D. S., Dash, P., Banyard, A. C., Barrett T. (2008). Full genome sequences of two virulent strains of peste-des-petits ruminants virus, the Cote d'Ivoire 1989 and Nigeria 1976 strains, *Virus Research, 136,* 192–197.

Chauhan, H. C., Chandel, B. S., Kher, H. N., Dadawala, A. I., Agrawal, S. M. (2009). Pesti des petits ruminants virus infection in animals, *Vet World, 2 (4),* 150-155.

Couacy-Hymann, E., Roger, F., Hurard, C., Guillou, J. P., Libeau, G. and Diallo, A. (2002). Rapid and sensitive detection of peste des petits ruminants virusby a polymerase chain reaction assay. *J Virol Methods, 100,* 17-25.

Dagalp, S. B., Sait, A., Faraji, A. R., Altay, H. M., Ozkul, A. (2015). Peste des petits ruminants in Turkey: findings between 2010 and 2013, *FAO-EMPRES Animal Health 360* No: 45/2015.

Dhar, P., Sreenivasa, B. P., Barrett, T., Corteyn, M., Singh, R. P., Bandyopadhyay, S. K. (2002). Recent epidemiology of peste des petits ruminants virus (PPRV). *Vet. Microbiol., 88 (2),* 153-159.

Diallo, A. (1988). Rinderpest and peste des petits ruminants; constant threats to animal farming in many developing countries. *Impact Sci. Society., 150,* 179- 192.

Diallo, A., Libeau, G., Couacy-Hymann, E., Barbron, M. (1995). Recent developments in the diagnosis of rinderpest and peste des petits ruminants. *Vet. Microbiol., 44*, 307-317.

Diallo, A. (2003). Control of peste des petits ruminants virus: Classical and New Generation Vaccines. *Dev. Biol. (Basel), 114,* 113-119.

Diallo. A. (2006). Control of Peste des Petits Ruminants and Poverty Alleviation. *J Vet Med., B53,* 11-13.

Diop, M., Sarr, J. and Libeau, G. (2005). Evaluation of novel diagnostic tools for peste des petits ruminants virus in naturally infected goat herds. *Epidemiol Infect., 133,*711-717.

Ezeibe, M. C., Okoroafor, O. N., Ngene, A. A., Eze, J. I., Eze, I. C. & Ugonabo, J. A. (2008). Persistent detection of peste de petits ruminants antigen in the faeces of recovered goats. *Trop Anim Health Pro, 40,* 517–519.

FAO, (1999). *Recognizing Peste des Petits Ruminants, A Field Manual.* http://www.fao.org/docrep/003/x1703e/x1703e00.htm.

FAO, (2009). *EMPRES Transboundary Animal Diseases Bulletin.* No: 33.

Forsyth, M. A. & Barrett, T. (1995). Evaluation of polymerase chain reaction for the detection and characterisation of rinderpest and peste des petits ruminants viruses for epidemiological studies. *Virus Research, 39*, 151-163.

Gargadennec, L. & Lalanne, A. (1942). Peste des petits ruminants. *Bull Services Zootechn Epizoot Afric Occident (French with English abstract) 5,* 16-21.

Gibbs, E. P. J., Taylor, W. P., Lawman, M. J. P., Bryant, J. (1979). Classification of peste des petits ruminants virus as the fourth member of the genus morbillivirus. *Intervirology. 11*, 268-274.

Gopilo, A. (2005). *Epidemiology of Peste des Petits Ruminants virus in Ethiopia and molecular studies on virulence.* PhD thesis. Institut National Polytechnique de Toulouse, France.

Gulyaz, V., Celen, N., Ozkul, A. (2005). Koyun ve keçi vebası (peste ses petits ruminants, ppr virüs izolasyonu, patojenitesi ve vero hücre kültürlerinde attenüasyonu çalışmaları. [Peste des petits ruminants, PPR virus isolation, pathogenicity and attenuation studies in vero cell cultures.] *Pendik Vet. Mik. Derg.* 36(1-2).

Gur, S., & Albayrak, H. (2010). Seroprevalance of peste des petits ruminants (PPR) in goitered gazelle (Gazella subgutturosa subgutturosa) in Turkey. *J Wildlife Dis., 46 (2),* 673-677.

Hamdy, F. W., Dardiri, A. H. & Nduaka, O. (1976). Etiology of stomatitis-pneumoenteritis complex in Nigerian dwarf goats. *Can J Comp Med., 40,* 276-284.

Haroun, M., Hajer, I., Mukhtar, I. and Ali, B. E. (2002). Detection of antibodies against peste des petits ruminants virus in sera of cattle, camels, sheep and goats in Sudan. *Vet Res Commun, 26,* 537-541.

Kwiatek, O., Ali, Y. H., Saeed, I. K., Khalafalla, A. I., Mohamed, O. I., Obeida, A. A., Abdelrahman, M. B., Osman, H. M., Taha, K. M., Abbas, Z., El Harrak, M., Lhor, Y., Diallo, A., Lancelot, R., Albina, E. and Libeau, G. (2011) Asian lineage of peste des petits ruminants virus. Africa. *Emerg Infect Dis., 17,* 1223–1231.

Lefevre, P. J. & Diallo, A. (1990). Peste des petits ruminants. *Rev. Sci. Tech. Off. Int. Epiz. 9 (4),* 951-965.

Lefevre, P. C., Diallo, A., Schenkel, F., Hussein, S. & Staak, G. (1991). Serological evidence of peste des petits ruminants in Jordan. *Vet Record*, 128, 110.

Libeau, G., Prehaud, C., Lancelot, R., Colas, F., Guerre, L., Bishop, D. H. and Diallo, A. (1995) Development of a competitive ELISA for detecting antibodies to the peste des petits ruminants virus using a recombinant nucleoprotein. *Res. Vet. Sci., 58,* 50–55.

Muthuchelvan, D., Sanyal, A., Balamurugan, V., Dhar, P., Bandyopadhyay, S. K. (2006). Sequence Analysis of the Nucleoprotein Gene of Asian Lineage Peste des Petits Ruminants Vaccine Virus. *Vet Res Commun, 30:* 957-963.

Nanda, S. K. & Baron, M. D. (2006). Rinderpest virus blocks type I and type II interferon action: role of structural and nonstructural proteins. *J Virol, 80 (15),* 7555-7568.

Nanda, Y. P., Chatterjee, A., Purohit, A. K., Diallo, A., Innui, K., Sharma, R. N., Libeau, G., Thevasagayam, J. A., Bruning, A., Kitching, R. P., Anderson, J., Barrett, T. and Taylor, W. P. (1996). The isolation of peste des petits ruminants virus from northern India. *Vet Microbiol., 51,* 207-216.

OIE, (2013). Peste des Petits ruminants, *Terrestrial Manual* 2013. Chapter 2.7.10. http://www.oie.int/fileadmin/Home/eng/Health_standards/tahm/2.07.10_PPR.pdf.

OIE, (2018). Peste des petits ruminants, Bulgaria. http://www.oie.int/wahis_2/public/wahid.php/Reviewreport/Review?reportid=27029.

Ozkul, A., Akca, Y., Alkan, F., Barrett, T., Karaoglu, T., Dagalp, S. B., Anderson, J., Yesilbag, K., Cokcalıskan, C., Gencay, A., Burgu, I. (2002). Prevelance, Distribution and Host Range of Peste des petits ruminants virus, Turkey. *Emerge. Infect Dis. 8 (7),* 708-712.

Reoder, P. L., G. Abrahim, et al. (1994). Peste des petits ruminants in Ethiopian Goats. *Trop Anim Health Pro, (26),* 69-73.

Reoder, P. L. & Obi, T. U. 1999. Recognising peste de petits ruminants: a field manual. *FAO Animal health manual no. 5. Rome, FAO Animal Health: 28.*

Roger, F., Guebre, Y. M., Libeau, G., Diallo, A., Yigezu, L. M., Yilma, T. (2001). Detection of antibodies of rinderpest and peste des petits ruminants viruses (*Paramyxoviridae*, Morbillivirus) during a new epizootic disease in Ethiopian camels (Camelus dromedarius). *Revue de Med. Vet., 152 (3),* 265-268.

SADC. (2012). *SADC Control Strategy for Pest de petits ruminants.* Gaborone, Southern African Developmenet Community: 1-24.

https://www.sadc.int/documents-publications/show/SADC%20Control %20Strategy%20for%20Peste%20de%20Petit%20Ruminants%20

Scott, G. R. (1981). Rinderpest andpeste des petits ruminants. In: *Virus Diseases of Food Animals,* Vol. II, E. P. J. Gibbs, (Ed.), Academic Press, New York, London, pp. 401-432.

Scott, G. R. (1990). Peste des petits ruminants (Goat plaque) virus. In: *Virus Infections of Ruminants.* Z. Dinter, B. Morein, (Eds). Elsevier Science Publisher, Amsterdam, Oxford, New York, Tokyo, chapter *33,* p. 355-361.

Sevik, M., & Sait, A. (2015). Genetic characterization of peste des petits ruminants virus, Turkey, 2009–2013. *Res. Vet. Sci., 101,* 187-195.

Shaila, M. S., Shamaki, D., Forsyth, M., Diallo, A., Goatley, L., Kitching, P., Barrett, T. (1996). Geographic distribution and epidemiology of peste des petits ruminants. *Virus Res. 43,* 149-153.

Wamwayi, H. M., Rossiter, P. B., Kariuki, D. P., Wafula, J. S., Barrett, T. & Anderson, J. (1995). Peste des petits ruminants antibodies in East Africa. *Vet Record, 136,* 199-200.

In: Advances in Health and Natural Sciences ISBN: 978-1-53614-639-4
Editors: Burcu Yuksel et al.

Chapter 16

ULTRA-HIGH MOLECULAR WEIGHT POLYETHYLENE FOR MEDICAL APPLICATIONS

Merve Dandan Doganci*[*] and Erdinc Doganci
Department of Chemistry, Kocaeli Vocational School,
Kocaeli University, Kocaeli, Turkey

ABSTRACT

Ultra-high molecular weight polyethylene (UHMWPE) is a subset of the engineering thermoplastic polymers with extraordinary mechanical and physical properties with a molecular weight of 2–6x10^6 $gmol^{-1}$. It is well known that UHMWPE has a great complex of convenient properties including thermal stability, wear resistance, and high impact strength, compared to customary polyethylene attained at low pressure. Additionally, UHMWPE has exceptional sound dampening characteristics, excellent electrical insulating properties, superior dielectric, chemical resistance, and low moisture permeability. All these inherent properties have enabled it to find widespread usage in various applications. This polymer has been extensively utilized in implants and

[*] Corresponding Author Email: merve.doganci@kocaeli.edu.tr.

orthopaedics as a biomaterial for total joint replacement applications since the 1960s. It enables effective tranfer load due to having extremely long chains. It is so flexible and light, has less weight, good impact strength when compared to metals. Recently, all these outstanding properties have enhanced the demand for this polymer for biomedical applications, including artificial hips, knees, elbows, as well as in shoulders, spinal disks, and wrists. The morphological structure and properties of UHMWPE and its applications for the manufacturing of medical devices were discussed in this chapter.

Keywords: implants, medical applications, total joint replacements, UHMWPE

INTRODUCTION

UHMWPE is a linear, low-pressure polyethylene homopolymer with a very long chain. It is one of the most important high performance polymers owing to extraordinary physical and mechanical properties: excellent impact and wear resistance, very low coefficient of friction, ease of manufacture, high impact toughness, outstanding dielectric properties, excellent aesthetics, noise and energy attenuation, good abrasion resistance, high resistance to cyclic fatigue, high fiber modulus and toughness, very low service temperature, resistance to radiation, and more (Hambir, 2000; Keller et al., 1994; Rein et al., 2007; Yilmaz et al., 2017; Živić et al., 2014).

Due to high impact strength and abrasion resistance properties of UHMWPE, it has the low friction coefficient which yields a self-lubricating nonstick surface. Besides, UHMWPE has a very low water absorption, high chemical and corrosion resistance, stability and biocompatibility in the body and it is one of the most available and cheap polymeric materials (Gabriel et al., 2010; Selyutin et al., 2010).

As the chemical structure and molecular architecture, UHMWPE is rather similar to high-density polyethylene (HDPE) differing in the average molecular chains lengths and molecular weights. Because it is produced by a process in suspension with Ziegler-Natta catalysts, which assigns the getting of plastics with molecular weight in the range of two and six million g/mol is 10–100 times higher than that of HDPE, it's called as "ultra-high" in linear polyethylene class (Haas et al., 2013; Kelly, 2002; Morais et al., 2017; Rein et al., 2007; Salles & Goncalves, 2003). The physical properties of UHMWPE in comparison to HDPE were summarized in Table 1.

Table 1. Typical physical properties of UHMWPE and HDPE (Fang et al., 2003; Kurtz, 2004)

Property	HDPE	UHMWPE
Molecular Weight (10^6 g/mol)	0.05-0.25	2-6
Melting Temperature (°C)	130-137	125-138
Poisson's Ratio	0.40	0.46
Specific Gravity	0.952-0.965	0.932-0.945
Tensile Modulus of Elasticity (GPa)	0.4-4.0	0.8-1.6
Tensile Yield Strength (MPa)	26-33	21-28
Tensile Ultimate Strength (MPa)	22-31	39-48
Tensile Ultimate Elongation (%)	10-1200	350-525
Degree of Crystallinity (%)	60-80	39-75

Owing to outstanding physical, chemical, and mechanical properties, it has been employed in many successful applications in different areas such as pulleys, textile, paper, the sliding parts of production lines, machinery, bumpers, mining, naval ropes, bearings, high-performance fiber products, artificial joints replacement in the human body (Forte & Ronca, 2017; Shafiee et al., 2014; Yilmaz et al., 2017).

UHMWPE FOR MEDICAL APPLICATIONS

UHMWPE is an ideal material for total joints with extremely large molecular chains which allow it to be used as a bearing surface and highly entangled chains make it resistant to wear. In 1962, Sir John Charnley used UHMWPE for joint articulation for the first time instead of polytetrafluoroethylene (PTFE) component because of its extensive wear resistance, lower friction and biocompatibility and it has been the crucial standard to produce one of the articulating surfaces of total knee, shoulder, and hip prostheses as well as elbow, ankle, total disc replacements (Musib, 2011). The main demands for a medical implant device in total joint replacements are low wear and friction, biological stability, biocompatibility, high toughness and creep resistance (Barbour et al., 1999).

To be used for biomedical applications, the resin powder must meet the essential properties as given in ISO 5834-1 and ASTM F648 standards (Kurtz, 2004). UHMWPE is initially fabricated as a micron sized resin powder (Fang et al., 2003; Musib, 2011). Later it is converted into consolidated form as a sheet, rod, or near shaped implant under high pressure and temperature by different thermoplastic-processing methods, including hot pressing, ram extrusion, sintering, compression moulding, spraying (hot flame, electrostatic) and gel formation for getting fibre (Figure 1). Because of the long polymer chain lengths, it has an enormously low melt flow index (MFI) or high melt viscosity of flow, 0.06 g/10 min, which restricts the processing of UHMWPE powder by conventional processing techniques, including screw extrusion, injection moulding, etc. (Gabriel et al., 2010). In the last step, it is machined from consolidated form into its final shape (Chilukoti & Periyasam, 2012; Kurtz, 2009).

The material must be sterilized before it's implanted into the human body. UHMWPE is generally sterilized with gamma irradiation in air or in a controlled environment with a sterilization dose of 25 and 40 kGy. The conventional sterilization techniques like autoclaving by steam cannot be used because the polymer begins to deform under high temperature conditions (Fang et al., 2003; Jones & Lauer, 1979). However, gamma sterilization can cause crosslinking, chain scission, and oxidation (Kurtz, 2004). The other methods for sterilization are ethylene oxide gas sterilization, low-temperature gas plasma sterilization, etc.

Figure 1. UHMWPE as a) resin powder form, b) sheet form, c) shaped form as a tablet.

The mechanical properties, oxidative degradation, and especially wear of UHMWPE are major problems for the longevity and performance of implants utilized in total joint replacements (Bracco et al., 2017; Fu et al., 2013; Fung et al., 2018; Jones & Lauer, 1979; Kurtz, 2004; Suñer et al., 2014; Wang et al., 1995; Wang et al., 1998). The bulk of the material is biocompatible, but particle debris due to wear are not biocompatible with the human body. It has been determined that millions of particles are taken apart from the surface of a total hip replacement (Kurtz et al., 1999). These wear debris can react with the tissue surrounding the bone, results in many problems for the patient such as cytotoxic reactions, inflammatory, and osteolysis (bone resorption). These reactions can cause loosening of the prosthesis. The improvement of the wear resistance could extend the clinical life span of total joint replacement and there has been ongoing research to solve the wear problem of UHMWPE (Affatoto et al., 2018; Baena et al., 2015; Holland et al., 2018). Oxidation is straightly correlated with the sterilization process of UHMWPE components. Sterilization with high-energy radiation causes high levels of oxidizing. Basically, high energy radiation creates free radicals which are highly reactive, and they easily react with oxygen causing oxidation on the backbone of the UHMWPE (Berry et al., 2012; Bracco et al., 2017). Besides, sterilization with gas or ethylene oxide (EtO) is the only method that effectively removes the oxidation (Affatoto et al., 2018; Del Prever et al., 2009; Mckellop et al., 2000). Oxidation causes to the formation of wear debris and makes changes in the mechanical properties of UHMWPE. Toughness of the UHMWPE composites is another important property for successful implementation of total joint replacements (Macuvele et al., 2017). High toughness restrains fracture and the creep resistance decreases the cold flow of the acetabular cup around the femoral head or into metal backings (Barbour et al., 1999).

CROSSLINKING OF UHMWPE

For nearly sixty years, UHMWPE has been commonly utilized as a bearing surface material in artificial bones and total joint replacements (Kurtz, 2009; Xiong et al., 2009). This semicrystalline polymer was selected for medical applications since it offers exclusive comprehensive properties like dynamic toughness, biocompatibility, high melt viscosity, high levels of strength, a low friction coefficient, creep resistance, and good wear resistance (Ansari et al., 2016; Macuvele et al., 2017; Pruitt, 2005). Owing to excessive adhesive/abrasive wear and resulted clinical complications like peri-prosthetic osteolysis are a major problem limiting the life span of implanted UHMWPE parts (Macuvele et al., 2017; Oral et al., 2009). Filling modifications for UHMWPE, like whisker, fiber, and particle reinforced UHMWPE composites have been researched to enhance its wear resistance rate (Xiong et al., 2009). Crosslinking of UHMWPE is an effective method to improve the wear resistance of polymer, to make it more durable and stable both in vitro and in vivo. First clinical research for crosslinked UHMWPEs demonstrated importantly a decrease in osteolysis and the wear rates, compared to original UHMWPE (Oral et al., 2008; Oral & Muratoglu, 2011). There are three classes of sterilization processes that have been utilized in crosslinking polymers such as chemical crosslinking using organic peroxides (chemical-induced crosslinking), ionizing radiation (radiation-induced crosslinking e.g., electron beam or γ-rays), and using silanes (silane compound-induced crosslinking) (Lewis, 2001; Muratoglu et al., 2001). Crosslinking is generated interchain covalent bonding of the long chains of the polymer by the reaction of the free radicals by these ways (Ansari et al., 2016). Despite crosslinking enhances the wear resistance, it can also lead to a decrease in the hardness, toughness, elongation at break, ultimate tensile strength, resistance to fatigue crack propagation, elastic modulus, and yield strength of the UHMWPE (Muratoglu et al., 2001). In 2000s, ionizing radiation crosslinked UHMWPE materials were introduced to clinical practice. This method forms free radicals and generates crosslinking mostly in the amorphous phase of the polyethylene while these free radicals formed throughout ionizing irradiation in the crystalline phase remained active for a long time (Oral et al., 2010). Residual free radicals are trapped into the amorphous phase which caused decomposition of mechanical properties and reduced the molecular weight of the UHMWPE in the long-term oxidative degradation with diffused oxygen (Fu et al., 2013; Oral et al., 2008; Oral et al., 2009).

Therefore, to mitigate the crucial problem, the crosslinked UHMWPE polymers were thermally treated (melted or annealed) after irradiation. Post-irradiation re-melting or annealing is used to enhance oxidation resistance in the first generation crosslinked UHMWPEs (Figure 2). Annealing method does not eliminate free radicals in long-term oxidation, but it stabilizes the mechanical and crystallinity properties of crosslinked UHMWPE. Post-irradiation re-melting as another approach decreases the free radicals, it also reduces fatigue strength of the radiation crosslinked UHMWPE owing to crystallinity loss during melting. The first generation of crosslinked materials was utilized to quench the free radicals which resulted in successful reducing wear, but reduced the crystal size, crystallinity, and the mechanical properties of these polymers like ductility, strength, and fracture toughness (Fu et al., 2013; Sobieraj & Rimnac, 2009). Accordingly, a second generation of crosslinked UHMWPEs enhances mechanical properties and oxidative stability, besides preserving the wear resistance of the first generation. Recently, different processing approaches including mechanical deformation, high pressure crystallization after melting highly crosslinked UHMWPE, solid-state deformation, 3-step annealing post-irradiation, and incorporation of vitamin-E have been proposed for the second generation (Puppulin et al., 2016; Sobieraj & Rimnac, 2009).

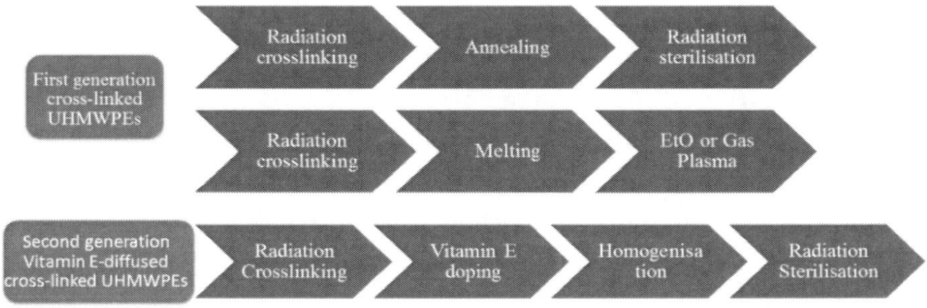

Figure 2. Schematic representation of processing of crosslinked UHMWPE (Oral & Muratoglu, 2011).

As an important alternative approach, free radical scavengers and/or antioxidants have been successfully utilized to promote the radiation crosslinked UHMWPE instead of post-irradiation melting (Oral & Muratoglu, 2011). Vitamin E (α-tocopherol) is widely used as a biocompatible, natural antioxidant agent in the medical fields and has been designed to enhance fatigue strength and mechanical properties of crosslinked UHMWPE compared to thermal treatments without sacrificing wear and oxidative stability (Oral et al., 2008). Because it can also act as an antioxidant during irradiation, it has successfully provided termination of free radicals to preserve long-term oxidative stability of UHMWPE. Conventional first and second generation crosslinking processes of UHMWPE are schematically depicted in Figure 2 (Oral & Muratoglu, 2011). Two methods of incorporate vitamin E in UHMWPE can be utilized to improve its wear resistance and mechanical properties; the first is to blend UHMWPE powders with vitamin E prior to irradiation and consolidation, and the second is to provide diffusion of vitamin E into molded and irradiated bulk UHMWPE blocks (Baena et al., 2015; Oral & Muratoglu, 2007; Oral & Muratoglu, 2011; Sobieraj & Rimnac, 2009).

TOTAL JOINT REPLACEMENTS

Joint replacement surgery takes out disturbed or diseased parts of a joint and repairs them partially or completely with new, synthetic parts called prosthesis. UHMWPE is the best material ever for most common artificial joints like hip and knee as well as shoulder, spinal, ankle or elbow prosthesis. UHMWPE parts of the joints are main load bearing surfaces that articulate against metal counterfaces like acetabular cup of the hip implants and tibial plateau of the knee implants. Each year, nearly 1.4 million or more joint replacement procedures are applied around the world. The hip and knee replacements are the most common joint replacements types.

Although ceramic-on-ceramic or metal-on-metal total joint replacements are experienced during the past decades, metal-on-UHMWPE is an international gold standard for damaged or diseased joints. UHMWPE coats the metallic surface of the joint to give a smooth surface and if the joints are not coated, friction can occur which is painful for the patients. Despite the enormous success and worldwide acceptance of this medical usage of UHMWPE, wear and resulting loosening problems limit the average life-span of the implants from 10-20 years as well as the specific conditions of each patient. There have been many efforts to solve this problem in UHMWPE at every stage of its productions from original to implanted forms (Baena et al., 2015; Kurtz et al., 1999; Li & Burstein, 1994). Especially, Kurtz and co-workers conducted numerous studies for the enhancement of UHMWPE and to improve its usage in total joint implants (Kurtz et al., 1999; Kurtz, 2004; Kurtz, 2009; MacDonald et al., 2018). Hydroxyapatite (HA), which is the major component of the mineral part of the human bone as reinforcement agent, was also used to improve the UHMWPE properties for biomedical applications (Macuvele et al., 2017).

Total Hip Replacements

The hip is the largest weight-bearing joint which consists of a ball and socket where thighbone joins with the pelvis and it is also enclosed by ligaments, muscles and cartilage inside the body. When some parts of the hip are destroyed, the joint movement becomes rigid and here occur fractures in the cartilage. The cartilage cannot repair or replace itself like other tissues in the body. Therefore, once cartilage is damaged, it is gone forever, and replacement is needed. In total hip replacement, which is also called total hip arthroplasty, the disturbed cartilage and bone are removed and replaced with orthopedic implants. The general concept of an implant is a combined system consisting of an acetabular component, a femoral stem with a femoral head and a plastic liner which is made from UHMWPE fixed between the femoral stem and acetabular component (Burger et al., 2007).

Various materials are used in total hip replacement such as metals, ceramics and polymeric materials. As mentioned before, UHMWPE was first used in total hip replacements in 1962 as an alternative material to PTFE and later gamma irradiated UHMWPE was introduced in 1968 by Sir John Charnley. Initially, PTFE was used as an acetabular cup, but UHMWPE has been widely used for its low friction and high wear-resistance. At high temperatures or under high stress, UHMWPE is easily deformed, and the resulting wear problem limits the long-term performance of joint prosthesis. In late 1990s, highly crosslinked and thermally treated polyethylenes (HXLPEs) were clinically introduced for total hip replacements to enhance wear resistance of the UHMWPE liners (Kurtz et al., 2011; Kurtz et al., 2015; McKellop et al., 1999; Muratoglu et al., 1999). These UHMWPEs became the alternative bearing material and showed an extreme wear resistance on hip simulators compared with conventional UHMWPE. It offers design flexibility, including smaller sizes. However, its usage is new, its long-term performance is still being closely studied (Ghalme et al., 2016; Teeter et al., 2017; Yamamoto et al., 2017).

Knee Replacements

Total knee replacements (total knee arthroplasty) are the other most successful joint procedures. UHMWPE was first used in the 1960's in knee replacement surgery. It is reported by the Agency for Healthcare Research and Quality that 600.000 or more knee replacements are applied in the United States every year. UHMWPE is a crucial material for total knee replacements and there is no other bearing material instead of it in the knee (Kurtz, 2004).

Implant design and geometry also play significant role in preventing wear in knee arthroplasty. The design of knee implants is more complicated than hip implants. There are two basic total knee implants: The most common one is fixed- bearing knees where UHMWPE of the tibial component is fixed tightly to the metal component and the second one is mobile-bearing or rotating platform implants which UHMWPE insert can rotate short distance inside the metal tibial tray and facilitates movement of the insert relative to the tray. The mobile-bearing knee implant was designed to reduce contact stress on the bearing UHMWPE surface and resulting wear. Even though this design has advantages due to the potential for reduced polyethylene wear and component loosening compared to fixed-bearing knee, this has not been exactly proven by clinical studies yet (Aglietti et al., 2005; Bartel et al., 1986; Hernigou et al., 2018; Huang et al., 2007; Kim et al., 2018; McEwen et al., 2005; Utzschneider et al., 2009). In addition, the thickness of UHMWPE insert for tibial components is a major factor that effects the contact stress and it was reported that the thickness of UHMWPE layer in tibial components for knee replacement should be at least eight to ten millimeters (Bartel et al., 1986; Chakravarty et al., 2015).

Shoulder Replacement

Shoulder replacement (total shoulder arthroplasty) is less common than knee or hip replacements. The first shoulder replacement was performed in the 1950s to treat shoulder fractures and UHMWPE has been used in shoulder replacements since 1970s (Barret et al., 1987; Neer et al., 1982). Total shoulder arthroplasty consists of a humeral stem which is generally made of a metal and a glenoid component which is a concave shaped UHMWPE socket. UHMWPE is responsible for motion and load bearing. The glenoid part is generally fixed by bone cement (polymethyl methacrylate, PMMA) or uncemented press-fit fixation can be used, too (Van Vrekhem et al., 2018).

Several types of implant design and fixation of the glenoid component have been described. The round-backed, all UHMWPE components with peg fixation have been reported as clinically performing better than flat-backed, metal-backed, or keeled components (De Wilde et al., 2013). The thickness of UHMWPE of glenoid components varies from 3 to 15mm. It is reported that thicker components are subjected to reduce polyethylene wear and stress. The size of the glenoid part below 4mm is not suitable while the thickness of 6mm seems the optimal size for reducing wear whenever the soft tissue tension allows it (Harryman et al., 1995; Terrier et al., 2012).

Ankle Replacement

Total ankle replacement (total ankle arthroplasty) is a surgical procedure for the treatment of end-stage ankle osteoarthritis. Initial total ankle replacement with a ball and socket type implant on the basis of total hip replacement was designed by Lord and Marotte in 1970 (Kurtz, 2009; Jamal et al., 2014). The first generation of total ankle replacements consisted of two components, a concave UHMWPE articular surface on the tibia and a metal (usually cobalt-chrome alloy) convex surface on the taller component. In some of the designs, the tibial and taller components were reversed (Bonasia et al., 2010).

Second generation implants are consisting of a UHMWPE bearing surface combined with the talar component or tibial plate. In this type implants cement is not used. The third-generation implants composed of three parts with a third UHMWPE component as the mobile bearing meniscus which is constrained to a minimum (Kim et al., 2016). There have been many commercially available ankle implant types, such as cemented, uncemented types, and implant designs could be classified by implant properties and surgical technique (Jamal et al., 2014; Yu & Sheskier, 2014; Barg et al., 2015; Sopher et al., 2017). There is no obvious information about the thickness of UHMWPE that should be used in total ankle replacement. The optimal UHMWPE component should be strong and thin, without the risk of impairing bony strength at the bone–implant–interface (Espinosa & Klammer, 2010).

Elbow Replacement

The elbow is a complex joint consisting of three bones: Humerus is the upper part, ulna and radius are the lower parts of the hinge joint. In total elbow replacement (total elbow arthroplasty), an artificial hinge, which is made of UHMWPE and special metal alloys, is located into the joint so all the bones can move smoothly on each other. Metal and UHMWPE hinge joints the stems together and allows the artificial joint to bend (Prkić et al., 2016). UHMWPE also minimizes the wear at the contact between the implant parts. Artificial joints come in different sizes to fit patients of different sizes.

There are mainly two types of implants: Linked types which have a mechanism linking the humeral and ulnar components and unlinked types (Sanchez-Sotelo, 2011). Both types of design rely on a metal on UHMWPE bearing couple for function (Kurtz, 2004). The unlinked elbow implants consist of two metal stems which are not mechanically jointed. The stems slide smoothly on each other and UHMWPE is used for minimizing the wear at the contact between the implant parts. Linked implants are also comprised of two metal stems which are connected with a metal locking pin. The locking pin attaches the ends of stems that are covered with UHMWPE (Bachman & Cil, 2017).

Total Disc Replacement

Total disc replacement (total artificial disc) is a surgical method in which degenerated intervertebral discs in the spinal column are replaced with artificial discs, a mobile bearing prosthesis made of metal and plastic, in the cervical (upper) or lumbar (lower) spine. Disc replacement surgery is a newer procedure that aims to preserve the intervertebral disc height of the spinal column, enhances disc flexibility, eliminates or reduces pain from motion, improves the quality of life of the patient. Total disc replacements are performed to relieve chronic low back pain caused by compressed nerves, weakness, and other symptoms due to damaged or diseased discs. They have been utilized in the lumbar spine since the late 1980s and recently in the cervical spine (Kurtz et al., 2005; Vital & Boissière, 2014). Total disc replacement devises typically include a CoCrMo alloy (Cobalt Chrome Molybdenum) socket articulating against an UHMWPE ball. Owing to its tribological properties and high chemical reaction resistance in part, UHMWPE has been commonly used as a core between the metal plates in joint disc arthroplasty as a bearing surface. There are two designs such as the Prodisc-L (Spinal Solutions) and SB Charité III (Link Spine Group) that include these components for their implant (Eckold et al., 2015). These modern devices are produced with a γ-inert-sterilized UHMWPE polymer to enhance wear performance and oxidation resistance of the cores. Wear debris is a major problem due to its effect on both implant longevity and the biological reactions that can occur in vivo as with all artificial joints (Eckold et al., 2015, Veruva et al., 2014).

Surgical Suture (Non Absorbable)

UHMWPE fibers have been used in the fields of medical applications, including ligament repair, surgery and trauma of the spine, high-strength sutures, arthroscopic procedures, motion-preserving spinal applications etc. For these fibers, suture technology was one of the first applications, with the trade name of Dyneema Purity, in 2004 (Kurtz, 2009). Sutures made with UHWMPE fiber are non-absorbable which present an exciting medical bio-engineering challenge. In the other words, it is displayed for use in ligation and/or approximation of tissue like cardiovascular surgeries and allograft tissue for orthopedic surgical procedures (Hapa et al., 2012; Kurtz, 2009).

Newer high-strength sutures comprised either partially or totally of UHMWPE demonstrate increased biomechanical properties over those of absorbable monofilament polydioxanone or nonabsorbable braided polyester suture utilized conventionally in arthroscopic repairs (Hapa et al., 2012). Sutures comprised of UHMWPE fiber were advanced to enhanced handling properties, tensile strength, limit gap formation, and decrease suture breaking. These UHMWPE sutures have submaximal resistance to slippage but certainly higher load to failure strengths than traditional braided polyester sutures (Barber et al., 2006; Barber et al., 2009). The strength of UHMWPE sutures was investigated in a porcine cadaveric model and determined that self-adjusting, UHMWPE suture-comprising meniscal repair instruments were comparable to isolated UHMWPE-containing suture repairs on load testing (Barber et al., 2006). Especially, UHMWPE sutures in equipment have further enhanced the mechanical fixation properties of meniscus repair devices. In another study, it was reported that inside meniscus repair by using UHMWPE sutures supplied greater load to failure strength than other repair techniques that utilize traditional sutures (Barber et al., 2009).

References

Affatato, S., Bordini, B., Fagnano, C., Taddei, P., Tinti, A., Toni, A., (2002). Effects of the sterilisation method on the wear of UHMWPE acetabular cups tested in a hip joint simulator. *Biomater. 23*: 1439-1446.

Affatato, S., Ruggiero, A., Jaber, S. A., Merola, M., Bracco, P., (2018). Wear behaviours and oxidation effects on different UHMWPE acetabular cups using a hip joint simulator. *Materials. 11*: 433-446.

Aglietti, P., Baldini, A., Buzzi, R., Lup, D., De Luca, L., (2005). Comparison of mobile-bearing and fixed-bearing total knee arthroplasty: a prospective randomized study. *J. Arthroplasty, 20*: 145-153.

Ansari, F., Ries, M. D., Pruitt, L., (2016). Effect of processing, sterilization and crosslinking on UHMWPE fatigue fracture and fatigue wear mechanisms in joint arthroplasty. *J. Mech. Behav. Biomed. Mater. 53*: 329-340.

Bachman, D., Cil, A., (2017). Current concepts in elbow arthroplasty. *Efort Open Rev. 2*: 83-88.

Baena, J. C., Wu, J., Peng, Z., (2015). Wear performance of UHMWPE and reinforced UHMWPE composites in arthroplasty applications: a review. *Lubricants, 3*: 413-436.

Barber, F. A., Herbert, M. A., Beavis, R. C., (2009). Cyclic load and failure behavior of arthroscopic knots and high strength sutures. *Arthroscopy: J. Arthroscopic Relat. Surg. 25*: 192-199.

Barber, F. A., Herbert, M. A., Coons, D. A., Boothby, M. H., (2006). Sutures and suture anchors-update 2006. *Arthroscopy: J. Arthroscopic Relat. Surg. 22*: 1063-1069.

Barber, F. A., Herbert, M. A., Schroeder, F. A., Aziz-Jacobo, J., Sutker, M. J., (2009). Biomechanical testing of new meniscal repair techniques containing ultra high–molecular weight polyethylene suture. *Arthroscopy: J. Arthroscopic Relat. Surg. 25*: 959-967.

Barbour, P., Stone, M., Fisher, J., (1999). A study of the wear resistance of three types of clinically applied UHMWPE for total replacement hip prostheses. *Biomater. 20*: 2101-2106.

Barg, A., Wimmer, M. D., Wiewiorski, M., Wirtz, D. C., Pagenstert, G. I., Valderrabano, V., (2015). Total ankle replacement: indications, implant designs, and results. *Dtsch. Arztebl. Int. 112*: 177-184.

Barrett, W. P., Franklin, J., Jackins, S., Wyss, C., (1987). Total shoulder arthroplasty. *J. Bone Joint Surg. Am. 69*: 865-872.

Bartel, D. L., Bicknell, V., Wright, T., (1986). The effect of conformity, thickness, and material on stresses in ultra-high molecular weight components for total joint replacement. *J. Bone Joint Surg. Am. 68*: 1041-1051.

Berry, D. J., Currier, B. H., Mayor, M. B., Collier, J. P., (2012). Gamma-irradiation sterilization in an inert environment: a partial solution. *Clin. Orthop. Relat. Res. 470*: 1805-1813.

Bonasia, D. E., Dettoni, F., Femino, J. E., Phisitkul, P., Germano, M., Amendola, A., (2010). Total ankle replacement: why, when and how? *The Iowa Orthop. J. 30*: 119-130.

Bracco, P., Bellare, A., Bistolfi, A., Affatato, S., (2017). Ultra-high molecular weight polyethylene: Influence of the chemical, physical and mechanical properties on the wear behavior. A review. *Materials. 10*: 791-813.

Burger, N. D. L., De Vaal, P. L., Meyer, J. P., (2007). Failure analysis on retrieved ultra high molecular weight polyethylene (UHMWPE) acetabular cups. *Eng. Failure Anal. 14*: 1329-1345.

Chakravarty, R., Elmallah, R. D., Cherian, J. J., Kurtz, S. M., Mont, M. A., (2015). Polyethylene wear in knee arthroplasty. *J. Knee Surg*. 28: 370-375.

Chilukoti, G. R., Periyasam, A. P., (2012). Ultra high molecular weight polyethylene for medical applications. *Technische Textilien. 55*: 100-103.

De Wilde, L., Dayerizadeh, N., De Neve, F., Basamania, C., Van Tongel, A., (2013). Fully uncemented glenoid component in total shoulder arthroplasty. *J. Shoulder and Elbow Surg. 22*: 1-7.

Del Prever, E. M. B., Bistolfi, A., Bracco, P., Costa, L., (2009). UHMWPE for arthroplasty: past or future? *J. Orthop. Traumatol. 10*: 1-8.

Eckold, D., Dearn, K., Shepherd, D., (2015). The evolution of polymer wear debris from total disc arthroplasty. *Biotribology. 1*: 42-50.

Espinosa, N., Klammer, G., (2010). Treatment of ankle osteoarthritis: arthrodesis versus total ankle replacement. *Eur. J. Trauma Emerg. Surg. 36*: 525-535.

Fang, H. W., Hsu, S. M., Sengers, J. V. (2003). *Ultra-high molecular weight polyethylene wear particle effects on bioactivity*. College Park, University of Maryland.

Forte, G., Ronca, S., (2017). Synthesis of disentangled ultra-high molecular weight polyethylene: influence of reaction medium on material properties. *Int. J. Polym. Sci. 1*: 1-8.

Fu, J., Shen, J., Gao, G., Xu, Y., Hou, R., Cong, Y., Cheng, Y., (2013). Natural polyphenol-stabilised highly crosslinked UHMWPE with high mechanical properties and low wear for joint implants. *J. Mater. Chem. B. 1*: 4727-4735.

Fung, M., Bowsher, J. G., Van Citters, D. W., (2018). Variation of mechanical properties and oxidation with radiation dose and source in highly crosslinked remelted UHMWPE. *J. Mech. Behav. Biomed. Mater. 82*: 112-119.

Gabriel, M. C., Mendes, L. B., De Melo Carvalho, B., Pinheiro, L. A., Capochi, J. D. T., Kubaski, E. T., Cintho, O. M., (2010). High-energy mechanical milling of ultra-high molecular weight polyethylene (UHMWPE). *Mater. Sci. Forum. 660*: 325-328.

Ghalme, S. G., Mankar, A., Bhalerao, Y., (2016). Biomaterials in hip joint replacement. *Int. J. Sci. Eng. 4*: 113-125.

Haas, I., Hubner, C., Kretschmer, W. P., Kempe, R., (2013). A highly efficient titanium catalyst for the synthesis of ultrahigh-molecular-weight polyethylene (UHMWPE). *Chem.–A Eur. J. 19*: 9132-9136.

Hambir, S., Jog, J., (2000). Sintering of ultra high molecular weight polyethylene. *Bull. Mater. Sci. 23*: 221-226.

Hapa, O., Barber, F. A., Suner, G., Ozden, R., Davul, S., Bozdag, E., Sunbuloglu, E., (2012). Biomechanical comparison of tibial eminence fracture fixation with high-strength suture, EndoButton, and suture anchor. *Arthroscopy: J. Arthroscopic Relat. Surg. 28*: 681-687.

Harryman, D. T., Sidles, J. A., Harris, S. L., Lippitt, S. B., Matsen, F. A., (1995). The effect of articular conformity and the size of the humeral head component on laxity and motion after glenohumeral arthroplasty. A study in cadavera. *J. Bone Joint Surg. Am. 77*: 555-563.

Hernigou, P., Huys, M., Pariat, J., Roubineau, F., Lachaniette, C. H. F., Dubory, A., (2018). Comparison of fixed-bearing and mobile-bearing total knee arthroplasty after high tibial osteotomy. *Int. Orthop. 42*: 317-322.

Holland, M., Fleming, L., Walton, K., Cerquiglini, A., Hothi, H., Hart, A., Skinner, J., Bills, P. J., (2018). Characterisation of wear areas on UHMWPE total knee replacement prostheses through study of their areal surface topographical parameters. *Surf. Topogr. Metrol. Prop. 6*: 34006-34019.

Huang, C. H., Liau, J. J., Cheng, C. K., (2007). Fixed or mobile-bearing total knee arthroplasty. *J. Orthop. Surg. Res. 2*: 1-8.

Jamal, B., Halai, M., Pillai, A., (2014). The evolution of total ankle arthroplasty. *Orthop. Muscul Syst. 3*: 145-148.

Jones, W. R., Lauer, J. L., (1979). Infrared analysis of polyethylene wear specimens using attenuated total reflection spectroscopy. *NASA Tech. Memo. 1*: 1-11.

Keller, A., Hikosaka, M., Rastogi, S., Toda, A., Barham, P., Goldbeck-Wood, G., (1994). An approach to the formation and growth of new phases with application to polymer crystallization: effect of finite size, metastability, and Ostwald's rule of stages. *J. Mater. Sci. 29*: 2579-2604.

Kelly, J. M., (2002). Ultra-high molecular weight polyethylene. *J. Macromol. Sci., Polym. Rev. 42*: 355-371.

Kim, D. R., Choi, Y. S., Potter, H. G., Li, A. E., Chun, K. Y., Jung, Y. Y., Kim, J. S., Young, K. W., (2016). Total ankle arthroplasty: an imaging overview. *Korean J. Radiol. 17*: 413-423.

Kim, Y. H., Park, J. W., Kim, J. S., (2018). Comparison of high-flexion fixed-bearing and high-flexion mobile-bearing total knee arthroplasties-a prospective randomized study. *J. Arthroplasty, 33*: 130-135.

Kurtz, S. M. (2004). *The UHMWPE handbook: ultra-high molecular weight polyethylene in total joint replacement* (1st edition). London, Elsevier.

Kurtz, S. M. (2009). *UHMWPE biomaterials handbook: ultra high molecular weight polyethylene in total joint replacement and medical devices (2nd edition.).* Academic Press, New York.

Kurtz, S. M., Gawel, H. A., Patel, J. D., (2011). History and systematic review of wear and osteolysis outcomes for first-generation highly crosslinked polyethylene. *Clin. Orthop. Relat. Res. 469*: 2262-2277.

Kurtz, S. M., MacDonald, D. W., Mont, M. A., Parvizi, J., Malkani, A. L., Hozack, W., (2015). Retrieval analysis of sequentially annealed highly crosslinked polyethylene used in total hip arthroplasty. *Clin. Orthop. Relat. Res. 473*: 962-971.

Kurtz, S. M., Muratoglu, O. K., Evans, M., Edidin, A. A., (1999). Advances in the processing, sterilization, and crosslinking of ultra-high molecular weight polyethylene for total joint arthroplasty. *Biomater. 20*: 1659-1688.

Kurtz, S. M., Peloza, J., Siskey, R., Villarraga, M. L., (2005). Analysis of a retrieved polyethylene total disc replacement component. *The Spine J. 5*: 344-350.

Lewis, G., (2001). Properties of crosslinked ultra-high-molecular-weight polyethylene. *Biomater. 22*: 371-401.

Li, S., Burstein, A. H., (1994). Ultra-High Molecular Weight Polyethylene. The Material and Its Use in Total Joint Implants. *J. Bone Joint Surg. Am. 76:* 1080-1090.

MacDonald, D. W., Higgs, G. B., Chen, A. F., Malkani, A. L., Mont, M. A., Kurtz, S. M., (2018). Oxidation, damage mechanisms, and reasons for revision of sequentially annealed highly crosslinked polyethylene in total knee arthroplasty. *J. Arthroplasty, 33*: 1235-1241.

Macuvele, D. L. P., Nones, J., Matsinhe, J. V., Lima, M. M., Soares, C., Fiori, M. A., Riella, H. G., (2017). Advances in ultra high molecular weight polyethylene/hydroxyapatite composites for biomedical applications: A brief review. *Mater. Sci. and Eng. C. 76*: 1248-1262.

McEwen, H., Barnett, P., Bell, C., Farrar, R., Auger, D., Stone, M., Fisher, J., (2005). The influence of design, materials and kinematics on the in vitro wear of total knee replacements. *J. Biomech. 38*: 357-365.

McKellop, H., Shen, F. W., Lu, B., Campbell, P., Salovey, R., (1999). Development of an extremely wear-resistant ultra high molecular weight polythylene for total hip replacements. *J. Orthop. Res. 17*: 157-167.

Mckellop, H., Shen, F. W., Lu, B., Campbell, P., Salovey, R., (2000). Effect of sterilization method and other modifications on the wear resistance of acetabular cups made of ultra-high molecular weight polyethylene: a hip-simulator study. *J. Bone Joint Surg. Am. 82*: 1708-1725.

Morais, F. L. D. D., Medeiros, F. D. S., Silva, G. G., Rabello, M. S., Sousa, A. R. D., (2017). Photodegradation of UHMWPE filled with iron ore fine. *Mater. Res. 20*: 356-364.

Muratoglu, O. K., Bragdon, C. R., O'Connor, D. O., Jasty, M., Harris, W. H., Gul, R., McGarry, F., (1999). Unified wear model for highly crosslinked ultra-high molecular weight polyethylenes (UHMWPE). *Biomater. 20*: 1463-1470.

Muratoglu, O. K., Bragdon, C. R., O'Connor, D. O., Jasty, M., Harris, W. H., (2001). A novel method of cross-linking ultra-high-molecular-weight polyethylene to improve wear, reduce oxidation, and retain mechanical properties: recipient of the 1999 HAP Paul Award. *J.Arthroplasty*, *16*: 149-160.

Musib, M. K., (2011). A review of the history and role of UHMWPE as a component in total joint replacements. *Int. J. Biol. Eng. 1*: 6-10.

Neer, C., Watson, K., Stanton, F., (1982). Recent experience in total shoulder replacement. *J. Bone and Joint Surg. 64*: 319-337.

Oral, E., Beckos, C. A. G., Lozynsky, A. J., Malhi, A. S., Muratoglu, O. K., (2009). Improved resistance to wear and fatigue fracture in high pressure crystallized vitamin E-containing ultra-high molecular weight polyethylene. *Biomater*. 30: 1870-1880.

Oral, E., Ghali, B. W., Rowell, S. L., Micheli, B. R., Lozynsky, A. J., Muratoglu, O. K., (2010). A surface crosslinked UHMWPE stabilized by vitamin E with low wear and high fatigue strength. *Biomater*. 31: 7051-7060.

Oral, E., Malhi, A. S., Wannomae, K. K., Muratoglu, O. K., (2008). Highly cross-linked ultrahigh molecular weight polyethylene with improved fatigue resistance for total joint arthroplasty: recipient of the 2006 Hap Paul Award. *J.Arthroplasty*. 23: 1037-1044.

Oral, E., Muratoglu, O. K., (2007). Radiation cross-linking in ultra-high molecular weight polyethylene for orthopaedic applications. *Nucl. Instrum. Methods Phys. Res., Sect. B: Beam Interactions with Materials and Atoms. 265*: 18-22.

Oral, E., Muratoglu, O. K., (2011). Vitamin E diffused, highly crosslinked UHMWPE: a review. *Int. Orthop. 35*: 215-223.

Prkić, A., Van Bergen, C. J., The, B., Eygendaal, D., (2016). Total elbow arthroplasty is moving forward: Review on past, present and future. *World J. Orthop. 7*: 44-49.

Pruitt, L. (2005). Conventional and cross-linked polyethylene properties. In: Bellemans, J., Ries, J. M. D., Victor, J. M. K., *Total Knee Arthroplasty* (1st edition, 353-360). Berlin, Germany:Springer.

Puppulin, L., Della Negra, S., Sugano, N., Sbaizero, O., Pezzotti, G., (2016). Surface modifications induced by in-vitro wear and oxidation on γ-irradiated UHMWPE hip liners belonging to different commercial generations. *J. Mech. Behav. Biomed. Mater. 53*: 414-426.

Rein, D. M., Shavit-Hadar, L., Khalfin, R., Cohen, Y., Shuster, K., Zussman, E., (2007). Electrospinning of ultrahigh-molecular-weight polyethylene nanofibers. *J. Polym. Sci., Part B: Polym. Phys. 45*: 766-773.

Salles, J., Goncalves, M., (2003). Effects of machining parameters on surface quality of the ultra high molecular weight polyethylene (UHMWPE). *Matéria 8*: 1-10.

Sanchez-Sotelo, J., (2011). Total elbow arthroplasty. *The Open Orthop. J. 5*: 115-123.

Selyutin, G., Gavrilov, Y., Voskresenskaya, E., Zakharov, V., Nikitin, V., Poluboyarov, V., (2010). Composite materials based on ultra high molecular polyethylene: properties, application prospects. *Chem. Sustainable Dev. 18*: 301-314.

Shafiee, M., Ramazani, A., Bahrami, H., Kheradmand, A., (2014). Preparation of ultra high molecular weight polyethylene using Ziegler-Natta catalyst system: Optimization of parameters by response surface methodology. *Iran. J. Chem. Eng. 11*: 55-62.

Sobieraj, M., Rimnac, C., (2009). Ultra high molecular weight polyethylene: mechanics, morphology, and clinical behavior. *J. Mech. Behav. Biomed. Mater. 2*: 433-443.

Sopher, R. S., Amis, A. A., Calder, J. D., Jeffers, J. R., (2017). Total ankle replacement design and positioning affect implant-bone micromotion and bone strains. *Med. Eng. Phys. 42*: 80-90.

Suñer, S., Bladen, C. L., Gowland, N., Tipper, J. L., Emami, N., (2014). Investigation of wear and wear particles from a UHMWPE/multi-walled carbon nanotube nanocomposite for total joint replacements. *Wear. 317*: 163-169.

Teeter, M. G., Yuan, X., Somerville, L. E., MacDonald, S. J., McCalden, R. W., Naudie, D. D., (2017). Thirteen-year wear rate comparison of highly crosslinked and conventional polyethylene in total hip arthroplasty: long-term follow-up of a prospective randomized controlled trial. *Can. J. Surg. 60*: 212-216.

Terrier, A., Brighenti, V., Pioletti, D. P., Farron, A., (2012). Importance of polyethylene thickness in total shoulder arthroplasty: a finite element analysis. *Clin. Biomech. 27*: 443-448.

Utzschneider, S., Paulus, A., Datz, J. C., Schroeder, C., Sievers, B., Wegener, B., Jansson, V., (2009). Influence of design and bearing material on polyethylene wear particle generation in total knee replacement. *Acta Biomater*. 5: 2495-2502.

Van Vrekhem, S., Vloebergh, K., Asadian, M., Vercruysse, C., Declercq, H., Van Tongel, A., De Wilde, L., De Geyter, N., Morent, R., (2018). Improving the surface properties of an UHMWPE shoulder implant with an atmospheric pressure plasma jet. *Sci. Rep. 8*: 4720-4731.

Veruva, S. Y., Steinbeck, M. J., Toth, J., Alexander, D. D., Kurtz, S. M., (2014). Which design and biomaterial factors affect clinical wear performance of total disc replacements? A systematic review. *Clin. Orthop. Relat. Res. 472:* 3759-3769.

Vital, J. M., Boissière, L., (2014). Total disc replacement. *Orthop. Traumatol. Surg. Res. 100*: 1-14.

Wang, A., Essner, A., Polineni, V., Stark, C., Dumbleton, J., (1998). Lubrication and wear of ultra-high molecular weight polyethylene in total joint replacements. *Tribol. Int. 31:* 17-33.

Wang, A., Sun, D., Stark, C., Dumbleton, JH, (1995). Wear mechanisms of UHMWPE in total joint replacements. *Wear. 181*: 241-249.

Xiong, L., Xiong, D. S., Jin, J. S., (2009). Study on tribological properties of irradiated crosslinking UHMWPE nano-composite. *J. Bionic Eng. 6:* 7-13.

Yamamoto, K., Tateiwa, T., Takahashi, Y., (2017). Vitamin E-stabilized highly crosslinked polyethylenes: The role and effectiveness in total hip arthroplasty. *J. Orthop. Sci. 22*: 384-390.

Yilmaz, G., Ellingham, T., Turng, L. S., (2017). Improved processability and the processing-structure-properties relationship of ultra-high molecular weight polyethylene via supercritical nitrogen and carbon dioxide in injection molding. *Polymers. 10*: 36-51.

Yu, J., Sheskier, S., (2014). Total ankle replacement-evolution of the technology and future applications. *Bull. Hosp. Jt. Dis. 72*: 120-128.

Živić, F., Babić, M., Adamović, D., Pelemis, S., (2014). Friction coefficient of uhmwpe during dry reciprocating sliding. *Tribology in Industry. 36*: 281-286.

In: Advances in Health and Natural Sciences ISBN: 978-1-53614-639-4
Editors: Burcu Yuksel et al.

Chapter 17

NURSING CARE FOR THE SYMPTOMS OF CANCER TREATMENT

Aylin Aydin Sayilan*
Department of Nursing, Kirklareli University, Kirklareli, Turkey

ABSTRACT

Cancer is the most common disease nowadays and it has lots of side effects for patients. Nursing care for symptoms of the cancer is very important for patients. This chapter reviews the symptoms of cancer treatment and nursing care.

Symptoms of cancer include skin-related side effects, hematologic side effects, side effects related to the nervous system, respiratory system, cardiovascular system, gastrointestinal system and genitourinary system and psychoemic side effects.

Keywords: cancer, nursing care, physiological systems, symptoms

---

* Corresponding Author Email: aylinay_85@msn.com.

INTRODUCTION

Cancer is a complex disease caused by uncontrolled division and proliferation of cells under the influence of genetic and environmental conditions. Serious systemic side-effects are observed both in the disease period and in the treatment period; nursing care for the symptoms comes into prominence.

This article, first, discusses the side effects of the cancer and then lists what the nurses should do for the care for symptoms during the treatment period.

1. Skin Related Side Effects

1.1. Alopecia

Hair loss begins 2-3 weeks after the use of the medications, the hair grows again 4-6 weeks after the end of the treatment.

Nursing Care

- Hair loss and hair regrowth after the end of treatment should be explained to patients.
- Patients should be informed about various materials such as hats, caps, wigs, etc., which can be used to hide temporary hair loss.
- Paying attention to ensure that the material to be applied to the head should be non-irritating for the head skin.
- Short-cut hair can be suggested before chemotherapy, eye pencil, eyebrow pencil, fake eyelashes can be suggested; use of a soft brush, hair drying without using hair dryer can be suggested for the remaining hair.
- Wearing a cold weather headgear for 30 days before and after the treatment can be suggested (Kanan, 1993; Ozbas, 2003; Akdemir & Birol, 2004).

1.2. Redness of the Skin

Nursing Care

- The patient should be informed about the possibility of skin redness, pruritus, skin drying, dermatitis and the risk of staying too long under the sun.
- The possibility of discoloration of skin and nail, the use of different colored nail polishes to hide this condition and the impermanence of this condition.
- The nails should be cut short; cotton clothes should be preferred.
- The importance of both the maintaining of hepatic enzyme control for hepatitis and daily control of the skin and sclera should be explained to the patient (Ertem et. al., 2009; Walter et. al., 2015).

1.3. Pruritus

Nursing Care:

- The skin should be dry and clean.
- Cold compresses, antihistamines and corticosteroids may be used.
- It should be explained to the patient that the skin should not be scratched to not exacerbate itching and to protect the integrity of the skin (Wu & Lacouture, 2018).

1.4. Hypersensitivity

Nursing Care:

- Urticaria, erythema, hyperpigmentation and acne-like reaction are observed within 1-3 days following the drug administration. It most commonly develops due to methotrexate, fluorouracil and doxorubicin.
- The patient avoids direct sun exposure during chemotherapy application.

- Cortisone is given as a premedication according to the request of the doctor.
- The patient is supplemented in terms of liquid.
- The patient is monitored for 2 hours after the end of chemotherapy. The patient's blood pressure is measured; adrenaline is available to be administered (Aydemir & Birol, 2004).

1.5. Extravasation

Extravasation is the penetration of vesicant chemotherapeutic agents into the subcutaneous tissues via veins.

Nursing care:

- The procedure which is administered is explained to the patient in detail; the patient is informed that there is no pain while the medication is giving.
- The anatomic region of injection or catheterization is monitored for an abnormal symptom such as swelling, redness.
- The physiological saline solution is given into the vein rapidly. Meantime the patient is asked to say when there is any discomfort.
- Chemotherapeutic drugs, primarily vesicants are administered in accordance with the standards established by the institution.
- Medications which are suitable for the diameter of the vein are given at the recommended dose.
- The needle/catheter is frequently checked to see if it is in place.
- The pressure is monitored when drug injection is required, the operation is stopped if there is a resistance.
- The anatomic region of injection or catheterization is carefully monitored.
- The information about the drugs that cause local reactions such as erythema and urticaria was known.
- The anatomic site of injection or catheterization is constantly checked by taking into account the sensitivities which are observed at the entry site or along the vein.

- After drug infusion, 10-20 ml of physiological saline solution is given rapidly.
- If there is any doubt about the vessel patency, the infusion is stopped immediately.
- The administration of the medication is stopped, the infiltrated amount is determined, the remaining amount in the injection syringe is determined.
- The chemotherapy injector is replaced with a new injector (20 ml).
- The drug or blood remaining up to 20 mL in the IV set was attempted to aspirate.
- If the indication is known, the antidote is slowly administered through the needle or catheter (IV).
- If it is not possible to aspirate the vesicant drug from the set, the needle is removed. Local ethyl chloride is applied to this numb region; the antidote is injected subcutaneously with a 25-gauge syringe.
- Bandages are applied without pressing on the affected area. The extravasation site is immobilized and the arm is held high for three days. It is regularly monitored in terms of pain, erythema and stiffness.
- If the pain still persists after 72 hours, a plastic surgery consultation may be needed (Chanes et al., 2012).

1.6. Skin Irritation

The aim of care in skin irritation is to minimize the skin reactions by assessing the site of the treatment and to maintain the integrity of the skin to prevent the development of infection.

Nursing Care

- Keeping the area clean and dry.
- Not rubbing and scratching the treatment region.

- Lightly washing the area with warm water and non-perfumed soft soap, drying with a soft towel.
- Not using talc powder, lotion or deodorant because they increase irritation.
- Wearing sweat absorbing, soft and abundant clothes, wearing long sleeved dress to protect the region from extreme heat and cold, direct sunlight, wearing wide-edge hats and using sun screen creams are important (Tuna 2014).

2. Hematological Side Affects

2.1. Anemia

The aim is to monitor signs of anemia, to detect them early and to take necessary precautions.

Nursing Interventions:

- Anemia signs such as fatigue, dizziness, trembling or tachypnea should be monitored.
- If orthostatic hypotension dizziness is observed, the patient should be informed to move slowly, especially when lying down and getting up.
- Hemoglobin and hematocrit should be monitored at regular intervals.
- The patient may need blood transfusion and erythropoietin according to the request of the doctor (Sabol et. al., 2010).

2.2. Neutropenia

- The signs of infection (fever, redness, temperature, edema, pain, inflammation, etc.) should be monitored.

- The sites where infection is mostly observed such as skin and mucous membranes, oral cavity, respiratory system and urinary system should be monitored; patient's oral and body hygiene should be attached importance and be performed daily, the skin should be protected from trauma.
- Perianal region should be cleaned with warm water and soft soap.
- Nails are cut short and straight.
- Skin integrity should be protected; skin should be cleaned Daily.
- The patient, his/her family and health personnel should take care to wash hands.
- The rest and nutrition of the patients should be cared (Vioral & Wentley, 2015).

2.3. Thrombocytopenia

The aim is to observe and to detect bleeding signs early, to prevent trauma related bleeding, to assess bleeding and to control bleeding.

- The blood count of the patient, especially the platelet level should be monitored.
- The platelet count should be kept above 20000.
- Patients should be protected from trauma.
- Soft materials should be used for oral care.
- Aspirin and non-steroid anti-inflammatory drugs should not be used.
- IM injections should be avoided; if necessary, pressure should be applied for 3-5 minutes after the procedure.
- Skin should be evaluated for petechial-ecchymosis.
- Unusual bleedings should be reported.
- If necessary, blood or platelet transfusion should be performed according to the physician's request (Esencan & Merih, 2010).

3. Nervous System Related Side Effects

3.1. Cerebral Edema

It is important to diagnose the symptoms of cerebral edema in the treatment of primary brain tumors or brain metastases by radiotherapy. These symptoms are considered as headache, nausea, vomiting, visual disturbances, inadequacy in motor function, and changes in mental status. It is stated that regular monitoring of the life signs of the patient with cerebral edema, the application of the planned treatment in the direction of the physician's request, frequent controls of the patient's level of consciousness are important (Jha, 2003).

4. Respiratory System Related Side Effects

4.1. Radiation Pneumonia

It is known that the possibility of the development of radiation pneumonia is high 1-3 weeks after the radiotherapy administered to lungs. It is emphasized that the presence of signs of fever and dyspnea indicates pneumonia in patients. Nursing interventions applied in patient care include administering steroids and antibiotics according to the physician's request and ensuring bed rest of the patient in addition to the interventions applied the patient with a coughing problem (Yılmaz et. al., 2014).

4.2. Radiation Fibrosis

It is stated that treatment related radiation fibrosis 6-12 weeks after the completion of radiotherapy is a restrictive pulmonary disease. The determination of dyspnea as the first symptom and the application of nursing interventions related to it have been addressed (Giridhar et. al., 2015).

4.3. Coughing

It is important to define the frequency, severity and characteristics of coughing that develops as a side effect of radiotherapy. The aim of care is to provide an effective open airway and to prevent both the deterioration of airway functions and the development of infection.

Nursing Interventions:

- Positioning the patient with the purpose of ensuring and maintaining airway patency.
- The application of respiratory exercises.
- Providing oral care frequently.
- Frequent ventilation of the room.
- Giving training to patients to prevent the use of irritant substances such as cigarettes that exacerbate coughing (Harle et. al., 2012; Molassiotis et. al., 2017).

5. Cardiovascular System Related Side Effects

5.1. Cardiotoxicity

- It may develop in the patient depending on the use of doxorubicin. If the patient takes the medication with a total dose of <550 mg/m^2, the incidence of cardiomyopathy is 0.5%. But, if the patient takes the medication with a total dose of >600 mg/m^2, the incidence of cardiomyopathy increases up to 30%.
- Changes in ECG such as tachycardia, extrasystoles, changes in ST-T waves were observed within the first week of drug administration. Rhythm changes are reversible. Congestive heart failure becomes chronic if it results in chronic cardiomyopathy.
- Doxorubicin should be limited to the dose of 450 mg/m^2.
- Congestive heart failure should be treated by using diuretics, digoxin and other cardiotonic drugs (Vasiliadis et. al., 2014).

6. Gastrointestinal System Related Side Effects

6.1. Stomatitis

It usually develops 7-14 days after the administration of the chemotherapeutic agent. Doxorubicin causes reduction in saliva production. Oral epithelial exfoliation occurs. Oral cavity erythema edema, mouth dryness, lips irritation, tongue edema or infection, superficial inflammation that may cause white irregular spots, fever blister like lesions are observed.

Nursing Care:

- Good oral hygiene is provided before and during chemotherapy.
- Oral cavity is checked and evaluated every day.
- Appropriate oral care is taught to the patient.
- If there is no lesion, the teeth are brushed with a soft toothbrush in 4 hours, mucostatin and betadine are used to gargle.
- If there is a lesion, oral care is carried out with hydrogen peroxide solution every two hours, it is followed by oral care with physiological saline solution, after that Vaseline is applied on the lips.
- Fluid balance is maintained.
- It is recommended to avoid from acidic, spicy, minty, salty and savory foods.
- If the lips are dry, they are softened with Vaseline or vitamin E preparations (Farrington et. al., 2013; Araújo et. al., 2015).

6.2. Diarrhea

The aim is to prevent diarrhea, to meet the needs of the body with an adequate and balanced diet and to maintain normal weight.

Nursing Care:

- The daily number of defecations and the characteristics of defecations are identified and recorded.

- Intestinal habits are defined.
- The fluid intake of the patient is monitored, lost fluid is replaced.
- The number of feces is recorded and each feces is assessed in terms of bleeding.
- Low fibrous foods, cooked foods are included in the menu.
- Milk products are avoided; if there is no objection, banana, orange, potatoes are given to compensate loss of potassium.
- The amount of fat in the diet is reduced.
- Patients should sparingly and frequently eat foods having low sugar content which are not too hot or not too cold (Shaw & Taylor, 2012).

6.3. Constipation

It is quite common because of the opioids used for the control of pain. 50% of cancer patients experience constipation, 78% of these patients are terminal-stage cancer patients.

Nursing Care:

- Fibrous foods are recommended (vegetables, fruits, oats, etc.).
- Fluid intake increase (at least 2lt per day).
- Physical activity is recommended.
- When non-pharmacological practices are not beneficial, laxatives may be given to the patient according to the physician's request, but enema application may be objectionable in these patients (Yu & Ko, 2010).

6.4. Nausea-Vomiting

Nausea and vomiting usually occur within the first 6 hours after the treatment. The aim of the care is to maintain fluid and electrolyte balance, to provide adequate nutrition and to protect the oral mucosa.

Nursing Interventions:

- It is important to evaluate and record vomited material in terms of quantity, content and color.
- It is important to administer antiemetic drugs at least half an hour before meals and the administration of chemotherapeutic drugs.
- It is important to provide oral care frequently.
- It is important to monitor body weight, fluid intake and fluid output and electrolyte values.
- The patient's methods to cope with nausea-vomiting are found out.
- The physical environment is kept clean and odorless; the clothes of the patient are ensured to be comfortable.
- Before and during the treatment, a cold pad is placed on the eyes of the patient; liquid and food restriction is done 4-6 hours before the treatment.
- It is suggested that the patient eat gradually and frequently, chew well, and avoid taking liquid at meals (Haugland & DeVon, 2018).

6.5. Anorexia

There is an unwillingness to eat and as a result of that the body weight decreases in time.

Nursing Interventions:

- Nurses should plan the nutrition program of the patient in cooperation with a dietician.
- Nurses should encourage the patient to eat high calorie and protein foods.
- Nurses should recommend to eat gradually and frequently.
- Nurses should recommend to perform oral care after each meal.
- Nurses should eliminate unpleasant odors or odors that the patient does not like.
- Nurses should maintain feeding via enteral nutrition or intravenous nutrition when the patient cannot feed orally (Berry et. al., 2018).

6.6. Hepatotoxicity

Prolonged use of methotrexate causes damage to the liver. Preexisting liver disease increases the risk. Hepatic fibrosis, hepatitis and an increase in liver enzymes are observed in the patient. The patient's liver function tests should be checked before chemotherapy begins (Cai et al., 2014).

6.7. Dryness of the Mouth

Saliva production decreases as a result of the effect of radiotherapy on salivary gland and therefore dryness of the mouth may occur.

Nursing Interventions:

- Informing the patients and their families.
- Emphasizing the importance of oral care.
- Encouraging the patient to eat juicy foods with meals.
- Giving advice to the patient to avoid from the foods containing lemon, glycerin and alcohol.
- Using commercial solutions to eliminate dryness of the mouth.
- Humidifying the patient's room.
- Encouraging the patient to intake plenty of fluid (Riley et. al., 2017).

6.8. Orthodontic Problems and Tooth Decay

It is stated that orthodontic problems and tooth decay are widespread as a result of dryness of the mouth in the patients who are treated with radiotherapy.

Nursing Interventions:

- Nurses should emphasize the importance of dental examination of the patient by a dentist before the treatment.
- Nurses should give a training on daily fluoride applications on teeth.
- Nurses should restrict the fluid intake for at least 30 minutes after fluoride applications (Gallarreta et. al., 2013).

6.9. Osteoradionecrosis

Osteoradionecrosis is considered as a late and chronic side effect of radiotherapy, is known to be a side effect that can occur in situations in which jaw bone structure is affected as a result of both tooth decay and infection. In the literature, it is stated that hyperbaric oxygen therapy may play a role in the healing of the bone as well as antibiotic applications in the treatment of osteoradionecrosis.

Nursing Interventions:

- Maintenance and support of oral care.
- Guidance for regular dental examinations.
- Monitoring of the suitability and comfort of dental prostheses.
- Informing the patient to decrease oral irritations caused by the use of substances such as cigarettes/alcohol comprise the goals of the nursing care (Firdoose & Hasoon, 2018).

6.10. Esophagitis

Esophagitis is seen during radiotherapy applications involving the esophagus region. The patient may report a pain/difficulty during swallowing and swelling in the throat. It is emphasized that it may be serious enough to prevent the practice of radiotherapy.

The intake of sweet, soft and liquid nutrients containing high amount of calorie and protein should be controlled in the diet of the patient. It should also be kept in mind that analgesics given 1-2 hours before meals may be effective in relieving pain and in maintaining nutrition of the patient (Etiz et al., 2013).

7. Genitourinary System Related Side Effects

7.1. Renal Toxicity

It occurs secondarily due to the collapse of the renal tubules. It most commonly occurs due to methotrexate.

Nursing Interventions:

- Fluid balance is maintained.
- Blood urea nitrogen (BUN), serum creatinine levels are checked.
- Creatinine clearance is measured.
- Uric acid and electrolytes in blood serum are analyzed and evaluated (Lydon, 1989; Mallick et. al., 2015).

7.2. Cystitis

It is reported that cystitis seen as a result of bladder radiotherapy, appears with thamuria, a decrease in bladder capacity and nocturia. In the literature, it has been stated that hyperbaric oxygen therapy improves the oxygenation of the tissues by providing granulation tissue formation and vascular growth; it is useful in the treatment of cystitis depending on radiotherapy.

Nursing Interventions:

- Fluid balance is maintained.
- BUN, serum creatinine levels are checked.
- The creatinine clearance is measured.
- Uric acid and electrolytes in blood serum are analyzed and evaluated (Lydon, 1989; Mallick et. al., 2015).

8. Psychoemotional Side Effects

8.1. Fatigue

Fatigue is one of the most common problems in cancer patients and is related with the factors such as nausea, vomiting, diarrhea, pain, immobility, anemia and malnutrition.

Nursing Care:

- The care needs of the patient should be fulfilled in a short time.
- Selecting light exercises according to the type of the treatment may help increase the energy level of the patient.

- It is ensured that patients do important activities when they have a high energy level, but they do not do these activities in a way that cause them to feel more fatigued afterwards; patients are ensured to get adequate amount of nutrition according to the Daily Living Activities Scale (DLA).
- It is recommended to drink at least 8-10 glasses of water per day in order to be able to remove collapsed cells and residual products (Mitchell et. al., 2014).

8.2. Pain

According to the suggestion of the WHO

- Non-narcotic analgesics + adjuvant therapy (aspirin, paracetamol) for mild pain.
- Poor narcotics + nonopioid-adjuvant therapy (tolvin, dihydrocodeine, demoral) for moderate pain.
- Strong narcotics + nonopioid drugs + adjuvant therapy (morphine) for severe pain.

Nonpharmacological applications recommended for pain control include position change, massage, warm or cold application, warm baths, providing a quiet and calm environment, teaching relaxation techniques, radio, television, reading, listening to music and dreaming (Yamanaka, 2018).

8.3. Anxiety and Depression

The symptoms such as stress, restlessness, headache and anorexia can be seen in the patient due to the remission (relief of symptoms of the disease), relapse (recurrence) and exacerbation (exacerbation) periods in chemotherapy.

Nursing Care:

- Positive communication and cooperation with the patient and his/her family should be maintained.

- If the patient feels a loss of control, his/her control should be restored or the patient should be ensured to feel that he/she controls his/her environment; this situation encourages the patient to struggle against difficulties.
- Sufficient psychosocial support should be provided to the patient and his/her family.
- For this reason, social counselors should be in the oncology centers.
- Physicians, nurses and other health professionals who are interested in oncology should be raised awareness on psychosocial oncology.
- The social services needed by the patient and his/her family should be determined.
- When giving information about the diagnosis and treatment, the patient should be informed with clear and understandable information (Yoo et. al., 2009; Tsaras et. al., 2018).

REFERENCES

Akdemir, N. & Birol, L. (2004). *İç hastalıkları ve hemşirelik bakımı: Alopesi*. Ankara, Sistem Ofset. [*Internal medicine and nursing care*: *Alopecia*. Ankara Sistem Publishing].

Araújo, S. N., Luz, M. H., da Silva, G. R., Andrade, E. M., Nunes, L. C., Moura, R. O. (2015). Cancer patients with oral mucositis: challenges for nursing care. *Rev Lat Am Enfermagem, 23,* 267-74.

Berry, D. L., Blonquist, T., Nayak, M. M., Roper, K., Hilton, N., Lombard, H., Hester, A., Chiavacci, A., Meyers, S., McManus, K. (2018). Cancer anorexia and cachexia: screening in an ambulatory ınfusion service and nutrition consultation. *Clin J Oncol Nurs., 22,* 63-68.

Cai, Z., Yang, J., Shu, X., Xiong, X. (2014). Chemotherapy-associated hepatotoxicity in colorectal cancer. *J BUON, 19,* 350-6.

Chanes, D. C., da Luz Gonçalves Pedreira, M., de Gutiérrez, M. G. (2012). Antineoplastic agents extravasation from peripheral intravenous line in children: a simple strategy for a safer nursing care. *Eur J Oncol Nurs., 16*, 17-25.

Ertem, G., Kalkım, A., Bulut, S., Sevil, Ü. (2009). Radyoterapi alan hastaların evde bakım gereksinimleri ve yaşam kaliteleri. *Maltepe Üniversitesi Hemşirelik Bilim ve Sanatı Dergisi, 2,* 5-11. [Home care needs and quality of life for patients receiving radiotherapy. *Maltepe University Nursing Science and Art Journal, 2*, 5-11].

Etiz, D., Bayman, E., Akcay, M., Sahin, B., Bal, C. (2013). Dosimetric and clinical predictors of acute esophagitis in lung cancer patients in Turkey treated with radiotherapy. *Asian Pac J Cancer Prev., 14,* 4223-8.

Farrington, M., Cullen, L., Dawson, C. (2013). Evidence-based oral care for oral mucositis. *ORL Head Neck Nurs., 31,* :6-15.

Firdoose, N., Hasoon, U. (2018). Efficacious use of a calgigraf ag foam dressing in complete healing of a difficult-to-heal, long-standing ulcer of osteoradionecrosis. *Natl J Maxillofac Surg., 9,* 78-81. doi: 10.4103/njms.NJMS_69_17.

Gallarreta, F. W., Pieroni, K. A., Mantovani, C. P., Silva, F. W., Nelson-Filho, P., de Queiroz, A. M. (2013). Oral changes stemming from hemangioma of the tongue. *Pediatr Dent., 35,* E75-8.

Giridhar, P., Mallick, S., Rath, G. K., Julka, P. K. (2015). Radiation induced lung injury: prediction, assessment and management. *Asian Pac J Cancer Prev., 16*, 2613-7.

Harle, A. S., Blackhall, F. H., Smith, J. A., Molassiotis, A. (2012). Understanding cough and its management in lung cancer. *Curr Opin Support Palliat Care, 6*, 153-62.

Haugland, T., DeVon, H. A. (2018). Symptoms, psychosocial factors, and health-related quality of life in patients with neuroendocrine tumors: an ıntegrative review. *Cancer Nurs,* doi: 10.1097/NCC.0000000000000614.

Jha, S. K. (2003). Cerebral edema and its management. *Med J Armed Forces India, 59*, 326–331.

Kanan, N. (1993). Kemoterapi uygulamalarında hemşirenin rolü. *Hemşirelik Bülteni, 7,* 31-40. [The role of the nurse in chemotheraphy applications. *Nursing Bulletin, 7*, 31-40].

Lydon, J. (1989). Assessment of renal function in the patient receiving chemotherapy. *Cancer Nurs., 12,* 133-43.

Mallick, S., Madan, R., Julka, P. K., Rath, G. K. (2015). Radiation induced cystitis and proctitis - prediction, assessment and management. *Asian Pac J Cancer Prev., 16,* 5589-94.

Mitchell, S. A., Hoffman, A. J., Clark, J. C., DeGennaro, R. M., Poirier, P., Robinson, C. B., Weisbrod, B. L. (2014). Putting evidence into practice: an update of evidence-based interventions for cancer-related fatigue during and following treatment. *Clin J Oncol Nurs., 18*, 38-58.

Molassiotis, A., Smith, J. A., Mazzone, P., Blackhall, F., Irwin, R. S. (2017). Symptomatic treatment of cough among adult patients with lung cancer: chest guideline and expert panel report. *Chest, 151,* 861-874.

Özbaş, A. (2003). Radyoterapi sonucu gelişen semptomlarda bakım. *Hemşirelik Dergisi, 13,* 57-73. [Care of symptoms in radiotherapy. *Nursing Journal, 13*, 57-73].

Riley, P., Glenny, A. M., Hua, F., Worthington, H. V. (2017). Pharmacological interventions for preventing dry mouth and salivary gland dysfunction following radiotherapy. *Cochrane Database Syst Rev., 7,* CD012744. doi: 10.1002/14651858.CD012744.

Sabol, V. K., Resnick, B., Galik, E., Gruber-Baldini, A. L., Gonce Morton, P., Hicks, G. E. (2010). Anemia evaluation and management in nursing home residents. *West J Nurs Res., 32*, 447-61.

Shaw, C., Taylor, L. (2012). Treatment-related diarrhea in patients with cancer. *Clin J Oncol Nurs., 6,* 413-7.

Tsaras, K., Papathanasiou, I. V., Mitsi, D., Veneti, A., Kelesi, M., Zyga, S., Fradelos, E. C. (2018). Assessment of depression and anxiety in breast cancer patients: prevalence and associated factors. *Asian Pac J Cancer Prev., 19*, 1661-1669.

Tuna, R. (2014). Safe handling of antineoplastic drugs in oncology nursing. *Health and Nursing Administration Journal, 2*, 105-108.

Vasiliadis, I., Kolovou, G., Mikhailidis, D. P. (2014). Cardiotoxicity and cancer therapy. *Angiology, 65*, 369-71.

Vioral, A. N. & Wentley, D. (2015). Managing oncology neutropenia and sepsis in the intensive care unit. *Crit Care Nurs Q., 38,* 165-74.

Walter, F. M., Rubin, G., Bankhead, C., Morris, H. C., Hall, N., Mills, K., Dobson, C., Rintoul, R. C., Hamilton, W., Emery, J. (2015). Symptoms and other factors associated with time to diagnosis and stage of lung cancer: a prospective cohort study. *Br J Cancer, 31*, 6-13.

Wu, J., Lacouture, M. E. (2018). Pruritus associated with targeted anticancer therapies and their management. *Dermatol Clin.*, *36*, 315-324.

Yamanaka, M. (2018). A concept analysis of self-management of cancer pain. *Asia Pac J Oncol Nurs., 5,* 254-261. doi: 10.4103/apjon.apjon_17_18.

Yılmaz Esencan, M. & Doğan Merih, Y. (2010). Kemoterapide semptomların yönetimi. *Zeynep Kamil Tıp Bülteni, 41,* 35-38. [Management of symptoms in chemotherapy. *Zeynep Kamil Medicine Bulletin, 41*, 35-38.

Yilmaz, S., Adas, Y. G., Hicsonmez, A., Andrieu, M. N., Akyurek, S., Gokce, S. C. (2014). Evaluation of the radiation pneumonia development risk in lung cancer cases. *Asian Pac J Cancer Prev., 15,* 7371-5.

Yoo, M. S., Lee, H., Yoon, J. A. (2009). Effects of a cognitive-behavioral nursing intervention on anxiety and depression in women with breast cancer undergoing radiotherapy. *J Korean Acad Nurs., 39,* 157-65.

Yu, C. T., Ko, N. Y. (2010). Evidence-based nursing care for cancer patients with opioid-induced constipation. *Hu Li Za Zhi, 57*, 100-5.

About the Editors

Burcu Yuksel is a Lecturer in Kocaeli Vocational School of Health Services at Kocaeli University in Turkey. She holds a master's degree in Biochemistry from Dumlupinar University and a PhD in Biology from Kocaeli University. Her research interests include plant biochemistry, molecular biology and genetics.

Mustafa Sencer Karagul received his BSc and PhD from Istanbul University, Faculty of Veterinary Medicine, Department of Veterinary Microbiology, Turkey. He was a member of National Reference Laboratory for Brucellosis. He is currently an assistant professor at Kocaeli University, Vocational School of Equine Science, Turkey. He has been conducting training on biosafety and biosecurity. His research interests include Brucellosis, vaccine and diagnostic antigen production and zoonotic diseases.

INDEX

A

abortus, 117, 118, 121, 124, 127, 137, 252
acclimatization, 41, 44
acetaminophen, 74
acid, 99, 103, 106, 110, 121, 136, 142, 151, 156, 160, 164, 167, 168, 169, 174, 176, 177, 179, 180, 190, 263, 266, 313
acid soil, 156, 174
acidity, 111, 124, 157, 172
activity level, 55, 58, 61, 62
acute febrile illness, xvi, 115
acute infection, 248
adaptation, 31, 41, 49, 168
adverse effects, 21, 157, 168, 204, 205
adverse event, 57
agricultural economics, 98, 109
agriculture, xviii, 19, 35, 43, 97, 108, 109, 150, 157, 180, 211, 261
agroterrorism, 116
air pollutant parameters, xiv, 2, 4, 13
air pollution, xiii, xiv, 1, 2, 3, 4, 5, 14, 15
air quality, xiv, 2, 4, 8, 9, 13, 15, 36
aldehydes, 191
alfalfa, 170
algae, 209
alkaloids, 113, 150
alkylation, 218
allelochemical, 98, 100, 102, 104, 105, 106, 112, 113
aluminum, x, xiii, xvi, 155, 156, 161, 170, 171, 172, 173, 174, 175, 176, 177, 178, 179, 180, 181, 221
aluminum tolerance, 156, 172, 174, 175
aluminum toxicity, xiii, xvi, 155, 156, 170, 171, 172, 173, 174, 175, 176, 177, 178, 179, 181
ambient air temperature, 38
amputation, 83, 84, 86, 87, 94, 95
anemia, 304, 313
anesthesia, 70, 73, 80, 187, 225
animal behavior, 185
animal disease, 249
annealing, 281
anorexia, 314, 315
antibiotic, 253, 312
antibody, 252, 269
antigen, 123, 266, 270
antioxidant, 99, 101, 103, 107, 160, 162, 167, 175, 176, 178, 191, 209, 282
antitumor, 199, 201
anxiety, 25, 31, 317, 318
apex, 155, 157, 166, 177, 179
apnea, 23, 27, 29, 30

Arabidopsis thaliana, 105, 108, 112, 209, 220
arginine, 164, 179
aromatic rings, 99
arthritis, 21, 23, 116
arthroplasty, 284, 285, 286, 287, 288, 290, 291, 292, 293, 294, 295, 296, 297, 298
aspiration pneumonia, 73
assessment, xvi, 26, 27, 48, 50, 61, 84, 110, 152, 210, 218, 316, 317
asymmetry, 184, 196, 199, 200
atmosphere, 2, 3, 11, 36, 120, 123
attitudes, 234, 239, 240

B

bacteria, 98, 118, 120, 121, 123, 125, 129, 132, 246, 248, 250
bacterial infection, 266
bandwidth, 186, 188, 195
bariatric surgery, 18, 23, 24, 27, 28, 31, 32, 33, 34, 81
biocompatibility, 276, 278, 280
biological activity, 110, 257
biological sciences, vii, 110
biomedical applications, 276, 278, 283, 294
biomolecules, 185, 188, 189
biotechnology, xiii, 113
bioterrorism, 116, 248
biovar(s), xvi, 115, 117, 120, 122, 124, 134, 253
blood, xv, 26, 27, 29, 30, 40, 41, 60, 72, 85, 91, 120, 123, 130, 266, 302, 303, 304, 305, 313
bonding, 193, 197, 202, 281
bone, 53, 54, 55, 61, 62, 63, 117, 130, 140, 279, 283, 284, 286, 287, 297, 312
bone health, 53, 54, 55, 62
bone marrow, 130, 140
bone marrow transplant, 130
bowel, 27, 31, 68, 71
brain abnormalities, 197
brain lateralization, 185
brain tumor, 24, 306
branching, 105, 112, 157
Brucella, xviii, 115, 116, 117, 118, 119, 120, 121, 122, 123, 124, 125, 126, 128, 129, 131, 132, 133, 134, 135, 136, 137, 243, 248, 250, 253, 254, 255, 256, 257, 258, 259, 260
Brucella abortus, 117, 120, 121, 122, 125, 127, 132, 137, 251, 254, 256, 257
Brucella canis, 117, 119, 121, 122, 123, 126, 136, 249, 259
Brucella ceti, 119, 121, 122
Brucella inopinata, 122, 123, 135
Brucella melitensis, 115, 117, 118, 125, 132, 135, 254, 255, 256, 257
Brucella neotomae, 119, 121, 122, 126
Brucella ovis, 117, 119, 120, 121, 122, 123, 126, 127, 137, 253, 255, 260
Brucella papionis, 122, 124, 136
Brucella pinnipedialis, 119, 122
Brucella suis, 117, 119, 121, 122, 125, 132, 137, 253
Brucella. melitensis, 117, 119, 121, 122, 124, 126, 127, 128, 130, 137, 247, 251, 252, 253, 254, 255, 256
brucellosis, x, xiii, xvi, xviii, 115, 116, 117, 118, 119, 123, 125, 126, 127, 128, 129, 130, 131, 132, 133, 134, 135, 136, 137, 243, 244, 245, 246, 247, 248, 249, 250, 251, 252, 253, 254, 255, 256, 257, 258, 259, 260

C

cadmium, 175, 219, 220
calcium, 171, 173, 175, 176
calorie, 18, 24, 310, 312
cancer, xi, xiii, xiv, xix, 3, 17, 20, 79, 99, 299, 300, 309, 313, 315, 316, 317, 318

carbohydrates, 19, 69, 74, 10, 1605, 186, 189, 190
carbohydrate therapy, 68
carbon, 26, 36, 181, 297, 298
carbon dioxide, 26, 36, 181, 298
cardiovascular disease, xiv, 17, 19, 20
cardiovascular system, 40, 299
catheter, 27, 31, 71, 75, 302, 303
cattle, 116, 117, 118, 119, 122, 127, 249, 252, 254, 255, 264, 266, 268, 271
cell cycle, 105, 150, 166, 173, 217
cell death, xvii, 108, 156, 160, 161, 162, 163, 164, 165, 168, 171, 172, 173, 174, 175, 176, 177, 178, 179, 180, 210, 217, 221
cell division, 143, 147, 149, 151, 158, 159
central nervous system, 74, 140, 184, 197
cerebellum, xvii, 183, 184, 185, 186, 187, 189, 196, 198
cerebral edema, 306
cerebral hemisphere, 200
chemical, xvii, xix, 37, 43, 99, 103, 111, 112, 146, 148, 149, 156, 183, 186, 199, 275, 276, 277, 280, 288, 291
chemotherapeutic agent, 302, 308
chemotherapy, 300, 301, 302, 303, 308, 311, 314, 317, 318
children, xiv, 1, 3, 21, 22, 43, 57, 61, 63, 316
chlorophyll, 103, 212
chloroplast, 160
cholesterol, 26, 189, 190, 193
chromosomal abnormalities, 140, 141, 149
chromosome, 108, 146, 147, 148, 149, 150, 153, 159
classification, 52, 86, 87, 93, 94, 121
climate change, xiii, xv, 35, 36, 37, 41, 43, 44, 45, 46, 47, 48, 50
clinical interventions, 85
clinical symptoms, 121, 123, 130, 264
coagulopathy, 41
collaboration, 243
colon cleansing, 68
colorectal cancer, 315
combination therapy, 22
COMET, xiii, xvii, 204
comorbidity, 23
comparative analysis, xvii, 184
compliance, 80, 187, 233, 250
complications, xv, 18, 23, 26, 28, 29, 30, 32, 57, 66, 67, 68, 71, 73, 75, 76, 83, 84, 116, 117, 250, 280
compounds, 3, 98, 99, 100, 102, 103, 107, 110, 111, 112, 146, 156, 214
confidentiality, 27, 225, 228, 233, 234
conformity, 290, 292
consciousness, 39, 41, 225, 306
consent, 224, 225, 233, 234, 238
consolidation, 185, 282
constipation, 309, 318
consumption, 2, 17, 26, 72, 104, 126, 128, 266
contamination, 122, 125, 151, 204, 264
control group, 103, 143, 144, 147
control measures, 245, 247, 267
correlation, 168, 201, 235, 238
coughing, 28, 29, 264, 306, 307
coumarin, x, xiii, xvi, 97, 98, 99, 101, 102, 103, 104, 105, 106, 107, 108, 109, 110, 111, 112, 113, 114, 154
creatinine, 26, 313
crops, 140, 155, 161, 167, 204, 211
cross-sectional study, 226
cultivars, 162, 164, 165, 177
cultivation, 209
culture, xv, xvii, 51, 105, 109, 136, 161, 204, 215, 220, 221, 267
cypermethrin, xiii, xvi, 139, 140, 141, 142, 149, 150, 151, 152, 153
cytochrome, 156, 162, 163
cytoskeleton, 156, 159, 177
cytotoxicity, 98, 143, 152

D

dandruff, 265
data collection, 11
deaths, 21, 35, 37, 42, 46, 66
debridement, 86
decision makers, 14, 245
DNA repair, 209, 217, 221
DNA strand breaks, 207, 212
dosage, xvi, 139
double helix, 159
drought, 36, 180
drug treatment, 22
drugs, 19, 26, 31, 54, 69, 70, 74, 186, 250, 302, 305, 307, 310, 314, 317
ductility, 281
durability, 212
dyes, 207
dyslipidemia, 26
dyspnea, 306

E

earthquakes, 37
ecology, 111, 112, 134
economic consequences, 95
economic development, 47, 59
economic losses, 48, 155, 161
economics, 98, 109
ecosystem, 104, 133, 203, 204, 257
edema, 304, 306, 308, 316
education, 58, 59, 61, 67, 77, 80, 84, 85, 86, 91, 224, 227, 229, 234, 235, 238, 239
e-Health, 60
El Salvador, 44, 47
election, 172
electrolyte, 26, 30, 68, 73, 104, 309, 310
electrolyte imbalance, 30, 68, 73
electrophoresis, xvii, 166, 204, 205, 206, 207, 208, 211, 212, 213, 218, 219, 221
emotional state, 40, 42, 51
endocrine, 19, 23, 32, 53, 72
endocrine disorders, 19
energy, 17, 27, 47, 52, 56, 61, 74, 102, 120, 148, 149, 203, 204, 276, 279, 292, 313, 314
engineering, xviii, 103, 275, 289
enhanced recovery after surgery protocol, 65, 66, 79, 81
environmental conditions, 250, 300
environmental contamination, 151
enzyme, 100, 101, 102, 103, 105, 160, 162, 167, 173, 214, 267, 301
EPA, 153
epidemiology, 14, 95, 126, 127, 128, 129, 130, 132, 246, 256, 257, 258, 270, 271, 273
eradication, x, xviii, 127, 130, 135, 243, 244, 245, 246, 247, 248, 249, 251, 254, 255, 256, 257, 258, 259, 261, 269
ethanol, 99, 202, 207
ethylene, 168, 180, 278, 280
ethylene oxide, 278, 280
exposure, 3, 37, 39, 41, 43, 46, 47, 48, 125, 140, 152, 158, 159, 160, 161, 165, 166, 168, 169, 209, 213, 214, 215, 216, 217, 256, 301

F

fast track surgery, 66
fatty acids, 196
fever, xvi, xviii, 71, 115, 116, 118, 129, 133, 257, 261, 265, 269, 304, 306, 308
fibers, 148, 149, 288
fibrosis, 306, 311
first generation, 281, 286
fish, 128, 140, 200, 218
fluid, 11, 13, 30, 40, 45, 69, 72, 73, 74, 76, 123, 125, 309, 310, 311
fluorescence, 174, 206, 207
food chain, 98, 128, 203, 204

formation, xvi, 87, 89, 97, 100, 102, 105, 107, 108, 141, 146, 147, 148, 149, 150, 156, 172, 174, 195, 207, 211, 278, 280, 289, 293, 313
fragments, 142, 149, 204, 211
frequency distribution, 230
friction, 276, 278, 280, 283, 284
FT-IR, x, xiii, xvii, 183, 184, 186, 187, 188, 189, 190, 192, 196, 198
functional changes, 195, 198
functional imaging, 184
fungi, 98, 102, 103, 209, 212
fusion, 263

G

gametophyte, 152
gastrectomy, 24, 25, 78
gastroesophageal reflux, 21
gel, xvii, 165, 166, 204, 205, 211, 212, 218, 219, 221, 278
gene expression, 162, 165, 167
general anesthesia, 73
generalizability, 228
genes, 120, 135, 149, 209, 259, 264
genotoxic, x, xvii, 100, 113, 114, 150, 152, 153, 159, 170, 203, 204, 205, 209, 211, 212, 213, 215, 217, 218, 219
genotoxic effect, xvii, 100, 113, 114, 150, 152, 153, 204, 209, 212, 213, 215, 217, 218, 219
genotoxicity, 98, 113, 179, 204, 205, 215, 216, 217, 218, 221
genus, 115, 119, 120, 121, 124, 128, 132, 135, 136, 262, 263, 269, 271
geostatistical analysis, 2, 8, 16
germination, 100, 101, 102, 104, 105, 106, 107, 108, 109, 110, 112, 141, 146, 152, 178, 180, 212
global climate change, xiii, xv, 35, 36, 37, 41, 44, 45, 48
glucose, 30, 60, 91, 120, 166, 169
glucose oxidase, 169
glutamate, 104, 105, 178
goat(s), xviii, 116, 117, 118, 122, 125, 127, 247, 249, 251, 252, 255, 256, 261, 262, 263, 266, 267, 268, 269, 270, 271, 272, 273
gravity, 188, 213, 217
greenhouse gas, xv, 36

H

habitats, 98, 161
healing, xv, 29, 65, 66, 68, 69, 70, 74, 86, 94, 312, 316
health care, 20, 56, 57, 58, 60, 87, 225, 227, 228, 234, 240, 241
health care professionals, 87
health problems, xiv, xvi, 1, 3, 17, 20, 36, 44, 53, 115, 116
health services, 60, 61, 224, 232, 233, 234, 237, 238
heat exhaustion, 42, 43
heat stress, 35, 36, 37, 39, 40, 41, 44, 45, 46, 48, 49
heat stroke, 39, 40, 41, 43, 46, 47
heavy metals, xvii, 178, 204, 209, 219
hemisphere, xvii, 183, 184, 186, 187, 196
hemispheric asymmetry, 184
hemorrhage, 30, 85
hepatoma, 113
hepatotoxicity, 315
herbicide, 102, 107, 108, 151
high school diploma, 235
hip arthroplasty, 284, 294, 297, 298
hip replacement, 279, 284, 285, 286, 294
hippocampus, xvii, 183, 184, 185, 186, 187, 189, 196, 198, 199, 201, 202
hormones, 51, 100, 215

hospital, x, xiv, xv, xvii, 18, 27, 47, 65, 66, 67, 69, 71, 72, 75, 76, 80, 119, 223, 224, 225, 227, 228, 230, 231, 240, 241, 254
hospitalization, xv, 27, 65, 72, 73, 76, 87, 225
human body, 38, 277, 278, 279
Human Brucellosis, 128, 133, 135, 258
human health, xiv, 1, 3, 36, 49, 128, 205, 245
human right, 224, 227, 233, 234, 238
husbandry, xvi, 115, 116, 126, 247
Hyacinthus orientalis, xvi, 139, 140, 146
hydrogen, 162, 178, 193, 197, 202, 308
hydrolysis, 111
hydroxyapatite, 294
hygiene, 243, 246, 305, 308
hyperbaric oxygen therapy, 312, 313
hypertension, xiv, 17, 20, 25, 42, 60
hypothermia, 72, 80, 81

I

image analysis, 213, 219
immune response, 263
impact strength, xviii, 275, 276
implants, 275, 276, 279, 283, 284, 285, 286, 287, 292, 294
in vitro, 141, 150, 152, 172, 178, 215, 217, 218, 221, 280, 294
in vivo, 150, 217, 219, 221, 280, 288
India, 43, 48, 262, 263, 272, 316
infection, xvi, xviii, 27, 29, 71, 72, 73, 75, 84, 93, 115, 117, 121, 122, 125, 126, 127, 128, 129, 130, 131, 134, 135, 175, 243, 245, 246, 247, 248, 249, 250, 251, 252, 253, 255, 256, 262, 264, 265, 266, 268, 269, 303, 304, 305, 307, 308, 312
infertility, xiv, xvi, 17, 115, 116, 117
inhibition, xvi, 70, 97, 100, 102, 105, 108, 113, 148, 149, 152, 157, 168, 173, 178, 180
insecticide, xvi, 103, 139, 140, 142, 150, 152, 212
integrity, 31, 123, 164, 166, 173, 301, 303, 305
International Olympic Committee, 59
intervention, 24, 30, 32, 61, 62, 83, 105, 225, 318
intravenous fluids, 75, 80
invertebrates, 140, 184
ionizing radiation, 202, 281
ions, 111, 157, 158, 163, 166, 167
IR spectroscopy, xvii, 184, 185, 186, 188, 198, 199, 200, 201, 202
irradiation, 199, 278, 281, 282, 291
isolation, 37, 119, 136, 209, 264, 266, 267, 272

J

joints, 277, 283, 287, 288

K

knee arthroplasty, 285, 290, 291, 292, 293, 294
knees, 276, 285

L

laboratory tests, 253, 266
laxatives, 309
leakage, 31, 68
left hemisphere, xvii, 183, 184, 187, 189, 196
leishmaniasis, 220
leisure time, 19, 52, 59
lesions, 95, 106, 117, 125, 129, 265, 308
level of education, 224
limbic system, 185

lipid peroxidation, 103, 158, 160, 162, 186, 191, 220
lipids, 186, 188, 189, 190, 191, 192, 195, 196, 200, 201, 209
livestock, 134, 135, 211, 246, 256
long-term memory, 185
loss of consciousness, 39, 41
lung cancer, 3, 316, 317, 318
lymph node, 125, 249, 265, 266

M

malate dehydrogenase, 105
malnutrition, 24, 313
Malta fever, 115, 116, 133, 257
mammals, 116, 119, 121, 122, 134, 262
manufacturing, xvi, xix, 139, 276
materials, 30, 31, 39, 100, 107, 276, 281, 284, 294, 296, 300, 305
matrix, 102, 263
mechanical properties, 276, 277, 279, 281, 282, 291, 292, 295
medical applications, xi, 233, 275, 276, 277, 280, 288, 291
medical care, 66
medical science, xiii
medication, 22, 24, 302, 303, 307
medicine, 23, 93, 97, 99, 109, 117, 244, 315
meiosis, x, xvi, 139, 140, 141, 142, 146, 147, 148
melitensis, 117, 121, 124, 127, 137, 251, 253
mellitus, 19, 83, 84
membranes, 125, 126, 129, 174, 193, 195, 200, 201, 202, 247, 305
meristem, 150, 151, 155, 157
meta-analysis, 76, 78
metabolism, 23, 40, 52, 100, 101, 102, 105, 141, 156, 197
metabolites, 98, 100, 101, 103, 113
metals, xvii, 178, 204, 209, 219, 276, 284
metaphase, 145, 146
methanol, 111
methodology, 296
methylation, 166, 178
microbial community, 107, 112
microclimate, 48
Micrococcus melitensis, 118
micronucleus, 146, 147, 149, 218
microorganisms, 116, 120, 209, 214, 219, 250
Middle East, 127, 130, 131, 262, 263, 267
migration, xiv, 2, 4, 11, 36, 37, 43, 207, 208, 214, 216
mitochondria, 156, 160, 162, 163, 173, 174, 191, 202
mitosis, x, xvi, 102, 108, 114, 139, 140, 141, 142, 143, 146, 148, 149, 150, 151, 154, 159
mitotic index, 108, 143, 147, 149, 150, 159
molecular structure, 199
molecular weight, xiii, xviii, 28, 99, 103, 275, 276, 281, 290, 291, 292, 293, 294, 295, 296, 297, 298
molecules, 162, 163, 167, 186, 189, 191, 193, 195, 196
morbidity, 21, 23, 66, 86, 262
morbillivirus, x, xviii, 261, 262, 263, 267, 269, 271, 272
morphology, 109, 111, 120, 123, 153, 157, 297
mortality, 21, 23, 48, 66, 68, 75, 83, 117, 262, 265, 268
mortality rate, 68, 265
mucous membrane, 129, 305
muscles, 38, 52, 56, 75, 284
musculoskeletal system, 20, 56
mutagenic, x, 100, 139, 140, 205

N

nail polish, 301

nanofibers, 296
nanoparticles, 152
narcotic analgesics, 70, 314
narcotics, 314
National Academy of Sciences, 197
National Institute for Occupational Safety and Health, 35
National Institutes of Health, 54
NATO, 116
natural compound, 99, 108
natural disaster, 36, 37
nausea, 73, 80, 306, 310, 313
negative effects, 98, 108, 205
negative outcomes, 58
nervous system, 41, 74, 140, 184, 197, 299
neurodevelopmental disorders, 198
neutropenia, 318
nitric oxide, 73, 169
nitrogen, 3, 105, 212, 214, 298, 313
nitrogen compounds, 214
nitrogen fixation, 212
nocturia, 313
nodes, 125, 249, 265, 266
non-structural protein, 263
normal distribution, 227
nuclei, 108, 147, 148, 177, 205, 207, 208, 213
nucleic acid, 151, 152, 186, 189, 190, 193, 263, 266
nucleus, 145, 146, 147, 148, 204, 207, 211, 213
nursing, vii, xiii, xiv, xix, 18, 32, 77, 78, 79, 80, 81, 140, 224, 225, 229, 239, 299, 300, 302, 306, 312, 315, 316, 317, 318
nursing care, 302
nursing home, 140, 317
nutrients, xvii, 155, 312
nutrition, 21, 22, 26, 31, 67, 75, 98, 109, 305, 309, 310, 312, 314, 315

O

obesity, ix, xiii, xiv, 17, 18, 19, 20, 21, 22, 23, 24, 25, 26, 27, 30, 32, 33, 34
obesity surgery, ix, xiv, 17, 18, 25, 27, 30, 32, 33
occupational health, ix, 35, 36, 37, 42, 43, 45, 47, 48, 50, 152
olanzapine, 198
organelles, 102, 162, 195
orthostatic hypotension, 304
osteoporosis, 53, 55
overweight, 17, 19, 20
oxidation, 110, 121, 160, 278, 280, 281, 288, 289, 292, 295, 296
oxidative damage, 160, 176, 177
oxidative stress, 156, 162, 163, 167, 170, 171, 175, 177, 209, 215, 219, 220
oxygen, xvi, 26, 56, 72, 102, 111, 120, 155, 160, 162, 166, 167, 170, 172, 175, 177, 191, 193, 202, 209, 220, 280, 281, 312, 313

P

pain, 29, 30, 67, 70, 72, 74, 76, 288, 302, 303, 304, 309, 312, 313, 314, 318
pain management, 29
palpitations, 44
Paramyxoviridae, 262, 263, 272
partial thromboplastin time, 26
pathogenesis, 132, 191, 246
pathogens, 37, 103, 110, 121, 129, 244
pathway, 79, 148, 149, 163, 173, 191, 217
patient care, xv, 65, 66, 77, 306
patient rights, xvii, 223, 224, 225, 226, 227, 228, 229, 230, 231, 233, 234, 236, 238, 239, 240, 241
peripheral neuropathy, 85, 87, 92
permeability, xix, 100, 106, 163, 275

peroxidation, 104, 158, 160, 162, 186, 191, 220
personal shoes and insoles, 84
peste des petits ruminants virus, 261, 268, 269, 270, 271, 272, 273
pH, xvi, 106, 120, 124, 155, 156, 158, 160, 161, 172, 205, 206, 207, 208, 210, 216
pharmacology, 176
phenolic compounds, 99, 100, 103, 110, 112
phenotype, 255
phosphate, 166, 175, 190, 193
phosphatidylcholine, 158
physical activity, xiii, xv, 17, 18, 19, 22, 30, 31, 51, 52, 53, 57, 58, 59, 60, 61, 62, 63
physical and mechanical properties, 276, 291
physical properties, xviii, 192, 275, 277
physical therapist, 87
physiological systems, 299
physiology, 100, 101, 102, 111, 174, 217
pigs, 117, 119, 126, 128, 253, 264
plant growth, xvi, xvii, 97, 103, 155, 156, 168, 203, 215, 216
plaque, 273
plasma membrane, 156, 158, 159, 164, 166, 178
pneumonia, xviii, 71, 73, 123, 136, 261, 306, 318
podiatry, 84, 94
podology, 83, 84
point mutation, 218
pollen germination, 141
pollen tube, 141, 178
polyethylenes, 284, 295, 298
polymer chain, 278
polymerase chain reaction, 163, 267, 269, 270
polysaccharide, 252, 256
postmenopausal women, 54
potassium, 175, 309
pregnancy, 51, 53, 55, 56, 57, 58, 140, 252
premenopausal women, 54
primary brain tumor, 306
programmed cell death (PCD), xvii, 156, 160, 161, 162, 163, 164, 165, 168, 171, 172, 173, 174, 175, 176, 177, 178, 179, 180, 210, 217, 221
prosthesis, 279, 283, 284, 288
protective mechanisms, 211, 217
proteins, 74, 100, 132, 148, 149, 158, 162, 178, 186, 188, 189, 192, 195, 196, 200, 209, 263, 272
psychological stress, 49
public health, xiv, 17, 18, 52, 59, 61, 84, 243, 245

Q

quality of life, 18, 21, 22, 23, 32, 53, 60, 83, 288, 316
quantification, 204, 207, 267

R

radiation, 36, 37, 38, 39, 43, 46, 47, 200, 201, 202, 276, 279, 281, 282, 292, 306, 318
radiotherapy, 306, 311, 312, 313, 316, 317, 318
RB51, 251, 253, 256
reactive oxygen, 160, 170, 172, 175, 191, 202, 209, 220
reconstruction, 25
renal dysfunction, 74
repair, 206, 209, 211, 217, 218, 220, 221, 284, 288, 289, 290
repetitions, 55, 142, 185
researchers, vii, xiii, xiv, xvii, 35, 100, 107, 119, 127, 128, 147, 156, 157, 159, 163, 164, 166, 167, 229
resistance, xviii, 69, 70, 110, 166, 171, 174, 178, 179, 217, 249, 252, 253, 275, 276,

278, 279, 280, 281, 282, 284, 288, 289, 290, 295, 302
resource utilization, 78
resources, 56, 128, 132, 247, 251
reticulum, 163, 171
Rev1, 251, 252, 253
rhabdomyolysis, 41
risk, xiii, xv, xvi, 21, 22, 23, 26, 27, 28, 35, 41, 42, 45, 50, 53, 57, 63, 68, 72, 73, 83, 85, 86, 87, 89, 91, 92, 93, 99, 115, 116, 122, 125, 129, 197, 205, 210, 249, 251, 252, 254, 264, 267, 287, 301, 311, 318
room temperature, 187, 206
root hair, 157
root system, 108, 112
ROS, 160, 162, 163, 167, 176, 209

S

S19, 251, 252, 253
safety, 29, 31, 37, 42, 45, 62, 76, 136, 225
Saudi Arabia, 60, 127, 268
second generation, 281, 282
sedative drugs, 70
sedentary behavior, xv, 51
seed, 102, 104, 105, 106, 108, 177, 178, 257
seedlings, 104, 105, 106, 109, 113, 153, 168, 170, 171, 174, 176, 177, 178, 181, 212, 213, 218
sensitivity, xviii, 112, 121, 206, 208, 219, 224, 225, 226, 227, 228, 233, 235, 236, 239, 240
sepsis, 68, 75, 318
sheep, xviii, 116, 117, 119, 122, 127, 133, 247, 249, 251, 252, 255, 256, 258, 261, 262, 263, 266, 267, 268, 269, 271
side effects, xiii, xix, 56, 57, 58, 74, 250, 299, 300
signal transduction, 159, 173
signs, 31, 86, 88, 89, 249, 266, 304, 305, 306
sister chromatid exchange, 150
skin, 20, 27, 38, 85, 92, 125, 129, 140, 265, 299, 300, 301, 303, 305
sleep apnea, 23, 27, 29, 30
Small ruminant morbillivirus, 261, 262, 263
spatial analysis, 2, 13
spatial memory, 185
sphygmomanometer, 27
spleen, 125, 140, 249, 266
stress factors, 155, 204
stress response, 70, 178, 179, 221
substrates, 56, 164, 165
surgery, xiv, xv, 18, 21, 23, 24, 25, 27, 28, 29, 30, 31, 32, 33, 65, 66, 67, 68, 69, 70, 71, 72, 73, 74, 75, 77, 78, 79, 80, 81, 94, 240, 283, 285, 288, 303
surgical nurse, xv, xviii, 65, 66, 67, 77, 224, 226, 228, 239
surgical nursing, 77, 80, 224
surgical service, xiii, xvii, 223, 224, 239
surgical technique, 287
surveillance, 128, 130, 131, 133, 135, 244, 245, 246, 257, 267

T

telophase, 145, 146, 147
temperature, 35, 36, 37, 38, 39, 40, 41, 42, 44, 45, 46, 72, 75, 92, 98, 111, 120, 187, 206, 212, 265, 266, 276, 278, 304
temporal lobe, 185
tensile strength, 281, 289
test-and-slaughter, 243, 244, 246, 247
therapy, 22, 69, 72, 250, 312, 313, 314, 318
thermal stability, xviii, 275
thermal treatment, 282
thermoregulation, 38, 41
thromboembolism, 30, 69, 75
thrombosis, 28, 29
tissue, xvii, 18, 20, 27, 29, 41, 53, 55, 58, 75, 86, 107, 108, 125, 129, 158, 170,

171, 183, 184, 185, 186, 187, 188, 189, 190, 191, 196, 197, 199, 200, 201, 204, 208, 215, 221, 264, 266, 279, 286, 289, 313
tobacco, xvii, 165, 173, 178, 204, 209, 211, 212, 213, 218, 219, 220
total joint replacements, 276, 278, 279, 280, 283, 295, 297, 298
toxic effect, 37, 97, 102, 109, 168
toxicity, xiii, xvi, 99, 141, 150, 153, 155, 156, 157, 158, 159, 160, 165, 167, 168, 170, 171, 172, 173, 174, 175, 176, 177, 178, 179, 180, 181, 211, 220
training, 224, 228, 230, 231, 239, 240, 307, 311
transmission, xviii, 126, 127, 129, 131, 191, 203, 243, 253, 264, 265
trauma, 58, 59, 66, 85, 288, 305
treatment, xv, 18, 20, 21, 22, 23, 24, 26, 27, 54, 66, 84, 86, 88, 89, 92, 93, 95, 106, 107, 124, 140, 146, 164, 167, 220, 224, 231, 232, 233, 234, 237, 238, 249, 250, 252, 253, 259, 268, 286, 299, 300, 303, 306, 309, 310, 311, 312, 313, 315, 317
treatment methods, 20, 23, 26
tuberculosis, 258
typhoid fever, 118

U

UHMWPE, xiii, xviii, 275, 276, 277, 278, 279, 280, 281, 282, 283, 284, 285, 286, 287, 288, 289, 290, 291, 292, 293, 295, 296, 297, 298
ulcer, 83, 85, 87, 89, 90, 91, 92, 93, 94, 316
urinary retention, 71

V

vaccination, 125, 243, 244, 245, 246, 249, 251, 252, 253, 254, 255, 267, 269
vaccine, 244, 248, 251, 252, 253, 255, 256, 257, 263, 267, 272
vacuole, 164, 167
variations, 98, 185, 187, 189, 193, 195, 200, 201
vascular surgeon, 87
visualization, 15, 211
vitamin E, 282, 295, 308
vomiting, 73, 80, 306, 309, 310, 313

W

walking, xvi, 56, 84, 85
walking analysis, xvi, 84
water, xvii, 27, 36, 41, 98, 99, 104, 124, 128, 142, 150, 155, 157, 158, 187, 193, 204, 207, 213, 264, 266, 276, 304, 305, 314
weight loss, 22, 24, 60
wild animals, 247, 264
wildlife, xviii, 126, 261, 262, 265
working hours, 42, 44, 45
World Health Organization (WHO), 17, 18, 20, 53, 116, 130, 250
wound healing, 29, 68, 86

Y

young adults, 197

Z

zoonosis, x, 115, 119, 135, 256, 259

Handbook of Diabetes for General Practitioners

Author: Anil K. Mandal (Fulbright Scholar, Founder of Mandal Diabetes Research Foundation, Consultant in Nephrology Putnam Community Medical Center, Palatka, Florida, USA, and others)

Series: Endocrinology Research and Clinical Developments

Book Description: The *Handbook of Diabetes for General Practitioners* is written for doctors to apprise the fundamentals of diabetes including its recognition as an entity, the discovery of insulin for the treatment of diabetes, pathophysiology of diabetes, and the development and progression of diabetic complications and how to minimize or prevent the complications.

Softcover ISBN: 978-1-63482-821-5
Retail Price: $95

Diabetes Mellitus Research Advances

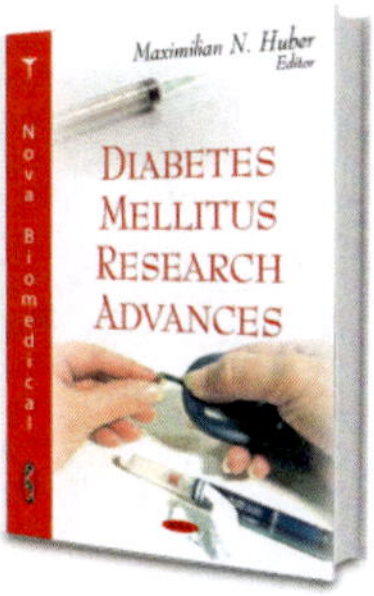

Editor: Maximilian N. Huber

Book Description: Diabetes mellitus is a chronic disease of absolute or relative insulin deficiency or resistance characterized by disturbances in carbohydrate, protein and fat metabolism. This new book presents the lastest research in the field.

Hardcover ISBN: 978-1-60021-711-1
Retail Price: $195